DARK DREAMS

DARK DREAMS

Alex Knight

Copyright © 2012 by Alex Knight.

ISBN: Softcover   978-1-4771-0734-8

All rights reserved. No part of this book may be reproduced or transmitted in any form or by any means, electronic or mechanical, including photocopying, recording, or by any information storage and retrieval system, without permission in writing from the copyright owner.

This book was printed in the United States of America.

To order additional copies of this book, contact:
Xlibris Corporation
1-888-795-4274
www.Xlibris.com
Orders@Xlibris.com

This book is dedicated to my mother,
"Anna Mae,"
for giving me strength and courage to
forge ahead in a world that can be very brutal.

INTRODUCTION

The book you are about to read is based on true research projects that are now taking place within all countries, the United States is no exception. I made a 1-800 call to a clinic within the U.S. to inquire about receiving the procedures of both stem cell implantation and being an embryonic donor, and was told "OK, just give me your name and number and I'll have one of our physicians call you." I asked "what state are you working from?" She replied "I can't give you that information, you'll have to ask one of our physicians." I just take names and numbers and forward them to a physician, and then they contact you." I asked "then why are you taking calls for these procedures in the U.S. if they aren't doing these procedures in this country." I could tell she was loosing her patients with me, and kept stating "I don't have any further information for you, just give me your name and number and I will give the information to a physician that does these procedures." I finished my questions hast fully before she could hang up. The questions I posed was "I would prefer to have these procedures done in my own country, can that be done?" Before hanging up she stated "you'll have to ask a physician, I just take names and numbers." Approximately within 3 months I attempted to renew my phone visit with the contact person for physicians continuing with these procedures. Not to my amazement in any way . . . . the phone number was—no longer a working number.

Stem Cells are cells from 3-5 day old embryos, fetus, or adult, under certain conditions and have the ability to reproduce itself for long periods of time. These embryos are fertilized outside of a woman's womb. Within the controlled laboratory scientists from all over the world are experimenting with embryonic stem cells, and with the understanding that there are specific guidelines that need to be followed. They are distinguished from other type cells by (2) characteristics. The first characteristic is that it can renew themselves every time they divide. The second characteristic of a stem cell is that under certain physiologic or experimental conditions, they can be included to become tissue or organ specific.

In 1998 a method to derive stem cells from human embryos and grown in labs are called Human Embryonic Stem Cells. These are the donated embryos with the consent of the donor and with specific guidelines.

In 2006 researchers could identify conditions that would allow some specialized adult cells to be reprogrammed genetically to assume a stem cell state. These embryos that are only 3-5 days old are called the blastocyt. Once removed from the blastocyt the inner cells mass can be cultured into embryonic stem cells which give rise to the entire body of an organ, e.g. cell type organs like the heart, lungs, liver and skin. These are known to be called "specific stem cells" which are created for a specific organ.

The embryonic stem cells which are fertilized outside the womb can be developed into many different types of cells in the body during their early life and growth of 3-5 days. In tissues they serve as an international repair system, and in muscle, they can repair the heart. They can also be used as general cells to replace red blood cells, and even brain cells.

An Embryonic Germ Cell Is derived from fetal tissue. They are isolated from the primordial germ cells. Primordial meaning that it came from the beginning, it was the origin, derived or developed as in the early Embryo formation. These in later development of 5-10 week fetus would have became the gonadal ridge and later develop into the testes or ovaries as the primordial germs cells can give rise to eggs or sperm.

Both Embryonic stem cells and germ cells are pluripotent are single cells and are not identical in their properties, embryonic stem cells and embryonic germ cells are not identical in their properties.

A Pluripotent stem cell, is a single stem cell which has the ability to give rise to types of cells that develop from the three germ layers,(mesoderm, endoderm, and ectoderm) from which all the cells of the human body arise. These pluripotent stem cells are those isolated and cultured from early human embryos and from fetal tissue that were going to be part of the gonads.

The hope to stop human suffering is the purpose of scientific research from stem cells. However there is much controversy about the use of human embryonic stem cells. There are efforts being made in a humanitarian manner for their use, and not as a dispute. It's about the legalities of the donor volunteers, and the guidelines which are involved. Past civilizations probably never imagined that advanced technology of medical science could create a way to sustain life without suffering.

There is no real way scientifically or technically to prove if an embryo before fertilization is considered to be a living person. So therefore, it is left for us to decide, each individual to use their own morals and ethics to provide judgment upon this research.

Ethics is not about knowledge, but about peoples morality issues and views. There is no right or wrong answers. It affects everyone in different ways. Stem Cell Research Ethics are opinions formed by information gathered from many different sources.

They were becoming more terrifying. Each and every time Bridgett Montgomery fell asleep, the dreams weaved their way into her REM sleep. (During REM sleep, the large voluntary muscles of the body are paralyzed. Conversely, brain activity is quite intense at this stage, increasing heart rate and blood pressure.) Now they were reaching deeper areas that were causing harmful medical conditions, as the rate of acceleration was becoming rapid and lasting longer. The longevity of the dreams seemed to be spiraling upward as unorganized columns of chromosomes. These dreams were now beginning to take over all sleep patterns, thus causing more serious consequences to occur. The heart palpations were becoming very painful, with stabbing pains clutching her chest tightly as if attempting to keep her heart in place. They continued escalating to the point of becoming unbearable and intolerable. The dreams presented themselves as a new form of horrific tortures, while forcing themselves upon Bridgett in many disgusting ways. The fears the dreams presented also brought feelings of destitution with the caption of being totally alone, losing self control, and being left with many insecurities which allowed no self control. Bridgett linked all these new emotions to ultimately dragging her mind into the darkest depths of Hell. She remembered before the onset of the project's development, starting with just a hypothesized revelation there appeared to be no endangerment concerning her well-being. As the projects unraveled research became heavily involved with the government, the health issues came into place as they took a huge decline downward. Bridgett had never been involved with anything so limitless. Faced with entities emerging with the onset of dreams and draining all energy sources, left her with the knowledge that something unknown and evil was awaking somewhere deep within her soul.

The source and origin of these dreams left her baffled, and she knew they would have to remain hidden. She was well aware that the dreams, if kindled, could cause monumental destruction to the project. Rumors were always rampant within her profession. It was a power play between scientists, and bets were placed upon whose projects would make it to the finish line. If the secret of the dreams were revealed, there would be irreconcilable damage regarding her sanity. Oh yes, Bridgett knew what they called scientists that went overboard with their projects—the "mad scientist syndrome." She would need to build a large block wall symbolizing a fortress within her mind to conceal them. Shielding was risky, and the reality of being anything of interest could be exposed reaching the "government's grapevine." There and any information given to this source would be used as a weapon of power, and this information was obtained by their greatest resources "snitches or spies." These individuals circulated within the lab making it difficult to decipher their identity. However all employees knew the motivation and drive behind them was collecting monetary benefits from the government. Bridgett's understanding was if a name was mentioned out of context and reported, the recourse would be the loss of the project and your job. This dent to anyone's profession in the field of science made it difficult to land another job with the government. Therefore, Bridgett's employment was dealing with an overly inquisitive, demanding, and very regimental employer who's constant reminder was that of control and power. These obscure attributes contained seizing anyone's research, stopping further funding, or shutting down a project at their desecration. It was obvious to Bridgett that there was no room for mistakes, as she was the closest scientist who held a theory that was almost completed. Bridgett held firmly to the many amazing secrets the project encompassed. Realizing that the dreams might interfere with the research of the project created a huge problem. She believed there was only a thin line between reality and abnormality. Of course, "abnormal" was another way of saying "crazy." "No," Bridgett scuffed, "I'll keep them hidden behind my secret wall beyond the sanctuary of my mind, locked tightly away from everyone. I will protect my project, and will take all measures insuring it's safety. The secrets of my dreams will never be revealed and will be taken to my grave." However courageous Bridgett portrayed herself to be, unanswered questions plagued her mind continuously. How dangerous are the dreams? Where did they come from? What do they want? These questions hounded Bridgett daily, even worse, whenever the dreams emerged.

Hanging on the walls of her personal bubble were Bridgett's most-prized possessions. She considered them to be tokens of the best years of her life, as summarized by all the accomplishments—BS in bimolecular engineering, MS in biology, and favoring the PhD on cell and developmental biology. Scanning through them with pride they unfolded a memory that encapsulated a question as to the obsession which motivated the desire to obtain all these achievements. Then the memories began to unfold bringing back crisp and happy thoughts dating back to childhood. Remembering the repetitive questions asked that took her mother to the edge and left noticeable puzzling expressions saying, "Bridgett, you have quite an imagination for a child your age. This would always symbolize she had no answers, as proved by sending her to seek out dad for these questions. Dad's answers were very straight and to the point. Bridgett acknowledged his words with intent detail of all explanations, echoing with clarity as if spoken yesterday. "You are the only person that is able to find the true answers you are seeking." Bridgett had forgotten the many wonderful memories that were being washed away in the bustling world, and wished there was more time to reminisce. Bridgett summarized its significance. "I guess the answer would be to find the past and stop the clock from ticking."

Bridgett was assisted by fellow scientists and colleagues whose degrees were also in biology and genetic; sharing many of the same interests. These professionals were already established as highly acclaimed scientists and believed in her project's theory of advanced medical and scientific technology. However, their assistance came with a payback system—a barter that was fully understood to be an agreement that any research provided would be exchanged for a job position. The scheme used had to contain accurate research supporting the theory in a positive manner. However, the clock kept ticking, and the sand in the hourglass was almost empty. The government was rearing its ugly head for proof of any advancement toward embryonic stem cell implantations. Bridgett had too much happening in life at this point and knew that the project's development was hanging by a thread. All government projects started as an experiment; however, this particular one would turn out to be the most expensive research ever funded. The reality of the research encompassing the project would involve it's entire developmental process. This drew the surreal observations of the many projects being shut down daily within the lab. Knowing, without a doubt, that these actions were either caused by a deficiency of funding or the lack of interest by the government. Tom, was Bridgett's colleague and confidante, and would often interject, "Remember what I told you Bridgett, if the government doesn't make enough money on a project, it will quickly and ultimately be erased." She was well aware that the government took no personal interest or consideration with any of their scientists. Destroyed research to them, was nothing more than just collateral damage.

The answer she was desperately seeking to continue with government funding came about by leaning toward another direction entirely. The new method would be achieved by using her innate ability of knowing people's characteristics and would keep the wolves at bay. However, now she would have to join their lair. Bridgett thought of herself as a sacrificial lamb taking part in the wolf pack; but this opened an innateness of a mother bear taking place. Bridgett understood there was a necessity to lower her standard of intellect. It was very crucial for the government officials to understand the research process of the project. This would be a slow process but would surface more clarity. Bridgett had succumbed to the realization that there had to be a definite bridge of communication extending from scientific data to the vocabulary of a layperson. This formula equated to the information exchanged from physician to patient; using simple wording that could easily be understood. It was imperative that this bridge was needed in the same fashion, but very cautiously. Bridgett perceived that this would allow the officials to understand the project's goal more clearly. After much deliberation, this conclusion would prove to be the most effective in leaving the line open for communication. At their level of intelligence this would provide a comfortable range and would benefit everyone. There was research deemed necessary to prove the theory of embryonic stem cell implantations; while succeeding in extending the longevity of mankind. Bridgett thoroughly understood the rules of engagement and the unpredictability of the government, with its sharp fangs containing toxic venom. Therefore, all inspections were never taken lightly, as their power determined the fate of the project.

Tom's explanation regarding the government's views of projects gave credence to Bridgett's belief; that with this truth there lied resentment between the project's reality, and the government's greed for money. She was being played as an instrument for the their amusement, and resented their ignorance. Her profession as a PhD on cell and developmental biology involved a project that would completely engulf her life; she would be forced to guard its intentions. One doubt that plagued Bridgett on a high scale was the true validity of the officials' credentials displayed on their badges. She felt a pinch of insult every time their laminated badges were flaunted, as the integrity of their credentials were very questionable. Did she want to argue about the validity of her credentials as opposed to theirs? Bridgett believed that displaying them in this obsessive manner was meant specifically to discredit hers, as their accreditations emphasized "government officials." They seemed to use them as taunting, and intimidating tools. Amused by the petty degrees of authority they enforced, and in defiance of justifying her own; Bridgett stood perfectly erect as the captain of the helm while research was gaining momentum.

The puppet masters seemed to enjoy the ambiance of being welcomed by Bridgett Montgomery. Yes, a new game began, so Bridgett could tolerate their existence. She definitely understood the power they entailed with their daily arrivals at the lab, and they made damn sure of it as well. The government's involvement with her particular project would be very lengthy because of the details ensuring its safety upon humans.

This factor alone would involve extensive funding. Bridgett understood without a doubt that one disagreement about the rules and regulations could result in the termination of the project. All questions asked by officials regarding the research project were answered. Bridgett would coil up like a snake waiting to strike; yet smart enough to know that their job was snake hunting, and it was apparent who would lose. The government had access to an onslaught of never-ending snake hunters; which were easily accessible as there was monetary gain involved. Therefore, she would respond merely as a domestic house pet holding back her feral predator abilities.

The officials' arrival each day brought an uncomfortable presence and was interpreted as an intrusion upon her space that left a bitter taste of being scrutinized. The continuation of questioning upon every step of the project's research introduced new feelings of interrogation. All these frustrations brought upon by the government's attacks were causing Bridgett to become slowly unglued. For some unknown reason there came about a new realistic character took over presenting itself as "The Phoenix" believed in many cultures to represent a mystical bird with its many colorful plumages of reds oranges and blues, depicted as the most beautiful bird ever created, bringing with it peace and harmony, rising up from the ashes. Bridgett still commenced with dodging the officials by hiding in different areas of the lab, and by portraying a feminine role of being quiet, shy, and docile that soon made the officials believe it as well; as told by other colleagues, also adding the implementation of being harmless. That was their first mistake. Bridgett would fit the profile of what the government believed her to represent—harmless. The new role portrayed for them was worthy of a Tony Award. She continued to welcome them with fake smiles while making them feel accepted. Bridgett's next "harmless" maneuver was to seduce the enemy by applying all of her God-given attributes toward the officials, as many had said that her physical attributes were found to be sexy and appealing. She pranced around the lab, being portrayed just that and nothing more. A harmless female scientist. The other scientists thought it was a hoot as they laughed in unison upon their breaks. The government officials didn't understand the bonds they all shared as they all belonged to MENSA. They didn't realize that all the crap they talked about her would soon come back to haunt them. Bridgett's colleagues offered up a lot of information regarding the officials. Therefore using these newly acquired skills were leading the officials off track and away from her personal space. She understood their job assignments. They were to obtain information from all the scientists, either through the understanding of the research papers or with answered questions. When finished with accomplished tasks, everything was handed over to the government. Bridgett began noticing the officials across the lab observing her every move. That's when the prancing commenced, as a trilogy of a Tony Award. They slowly began dissipating from the time spent around her area. Bridgett continued portraying this new feminine fatile mode, until she was told that they finally assumed her to be intimated quite easily. "That's a new one, Bridgett thought." "Who's intimidated by who?" Tom knew very surreal those traits associated to Bridgett and none equated to the thoughts entailed by the officials. He spared nothing when it came to her and announced quite frequently to many, "The whole package comes with, "beauty and brains," which equated to "quite a prize to win." Through the happiness of seducing the enemy, it was still difficult to shake the label "harmless." Bridgett sloughed off the other names attached; but never that one word. She toyed with that specific word usage and kept the thoughts hidden. Bridgett was compliant with all the rules and regulations and left the communication gate or bridge open at all times for the officials to understand the research progress of the project. However there came about another hurdle to jump over. The government began implementing stricter laws and guidelines. It didn't take a piece of a space shuttle to drop from the sky and land on her head to understand the reasons. Bridgett acknowledged that the assigned project owned by the government was now considered **top secret**. She also recognized that the one-way mirrors were a secret trick enabling them to observe her every move, ensuring that their new tactics were being enforced, and they weren't playing.

Bridgett became furious with the government for using their power in a reversed manner, instead of assisting the project's needs. She was pissed, and would now become devious and candid in allowing them to figure her out. Bridgett's personalities could switch like that of a chameleon just as fast as the government's implements of new procedures. She would now amplify her work at the lab as boring, causing an expression that symbolized a trait called the "flat affect," producing an I-don't-care attitude. This effect worked well with everyone she wanted out of her circle of friends as it introduced a facial expression that showed no physical emotions. Whenever the officials came close asking questions or requesting used research; she gritted her teeth while speaking to them. Tom always made comments about Bridgett's looks, saying she had very pretty pouty lips and a beautiful smile that could stop a war. With those attributes removed and replaced with a vacant look that expressed nothing, the new physical appearance proved to the officials that she was as they thought her to be: just another boring scientist, as well as with other titles that shouldn't be applied to a lady. They tried many ways to bring back the lab that Bridgett once filled with joy and laughter. The one with the wonderful persona. The other Bridgett, the one with the free spirit, sparkling personality, and friendly welcomes. Their attempts at everything possible to seek her out was failing. Their conversations and stupid jokes were not being accepted. Bridgett was now all work and no play. She was finished playing games and focused on the project, leaving a presence still exhibited as being cold as ice. It seemed that the rules of the game were reversing, as they took on roles as circus clowns, exploiting anything that would gain her attention, and were very zealous with them. There were times when she felt pity for their performances and would throw them a dog biscuit, which was a reused piece of research no longer important to the project. Bridgett finally saw a state of perplexity upon their faces each time the research papers were turned in to them. They no longer asked any questions regarding the project, and this was a large part of their job description. This amplified the confidence she needed by watching their uneasiness, and being unable to exert any power they enjoyed afflicting. The means to an end came when Bridgett realized that all the research used for the project given to the officials; per government regulations, left them clueless to the project's conclusive state. The answer she was seeking since the first day of their attack upon the lab, as to the validity of their credentials, gave proof of their incompetence.

Bridgett understood the reason she wasn't enjoying their games or felt the urge to play any longer. It was mental fatigue taking its toll. The stress of the government officials and the top priority placed upon the project had been a long and grueling three years. The onslaught of the roaches making homes in the lab was leading to a meltdown. It seemed as if the government was purposely hindering her ability to stay with the project. She was convinced that they were forcing a medical leave upon her as a way to relinquish the project for the sake of its existence to another scientist. Once again they were exploited as pushing their power and exerting different types of authority without a challenge. They were throwing everything possible her way, but she caught everything and threw it back at them. Maybe the infantile fun and games gave their pathetic lives a meaning to exist. Some officials, she thought, were using their jobs as a means of a power play with the scientists, while others had personalities of just pure disrespect and hatred. Their behavior was anything but entertaining or socializing especially when suppressing intelligence where communication skills were necessary. Their presence were causing her migraines. Bridgett knew that the chaotic environment in the lab had to be eliminated in order to alleviate the migraines. This meant that she would have to push herself into overdrive every day and night until the project reached a safe and conclusive end. This overtime would cause more sleep deprivation.

Sometimes at work, coming up for an exchange of fresh air, brought "flashbacks" of the lab prior to the government's abrupt arrival. She would recall and savor all the memories from when it once contained life. The laughter and socializing gave comfortable feelings of knowing that everyone had your back, and the wonderful smell of freshly brewed coffee was an added appeal. The chatter and mixture of conversations became a learning period of wealth and knowledge obtained from very prominent and well-acclaimed scientists. Their intellectual stimuli's were accelerating. They all shared common bonds that needed no explanation as they were all members of MENSA. This was a reason their bonds would never be broken, as it kept them within their own circle and they protected each other. Many scientists were introverted people, loners, and some chose to be disconnected from the rest of the world. Bridgett now understood the whole picture for their reasons, as her fondest memories were just that and nothing more. The environment within the lab started changing dramatically when the officials arrived, placing all their new polices, rules, and regulations into effect, while turning the lab into an invisible prison. How she hated the government for allowing the officials to turn all their good intentions of productivity into a forceful labor camp. The fake smiles portrayed, the dull jokes, and the lack of intellectual conversations that the officials presented were hardly tolerable. Their demands of power and dominance while attempting to achieve respect were totally absurd. "What the Hell was it all about?" Bridgett often questioned. "Did they really believe this type of strategy was appropriate? Did they think that scientists were nothing more than intelligent weapons to be used at their discretion.

The government official's presence was like dealing with parasites in a feeding frenzy of delight. The calculations of the approximate due date of all research had to be completed timely or beware of big brother: which was included as well. "Would I meet the government's deadline?" She needed to move quickly to the next chain of events and any new games that were on the horizon. Their primitive thoughts of devastation and disaster devoured her with their vigilance surrounding the project. Bridgett couldn't rid the huge dark cloud that seemed to plague the lab. She could only focus on all other unanswered inquires that involved her subconscious. Will today be the last day of my research? Would they find a way to remove and replace me with another scientist? The thoughts regarding secrecy and top priority of the project became a huge concern, as she compared it to that of the **Einstein effect**. This involved the government taking a project that was thoroughly researched to be safe and used their twisted ideas making it a powerful and destructive weapon instead. The conclusion, brought the answer that was dreaded. Bridgett was dealing with the devil, and any road taken was blocked. The question that stood out in very large print was, "What do they want with me?" The fact that she had been handpicked by the government left her puzzled. These twisted impositions Bridgett was forced to endure only projected menacing thoughts. Another great concern was whether or not she would meet the government's deadline, as emphasized by the funding involved with the project. Conversations with other scientists that had worked for the government came up with some unusual information. They stated that the government had a way of making scientists "paranoid" to the point of hospitalization. Bridgett was more determined then ever not to become a statistic.

Sleeping was interrupted constantly causing severe insomnia issues. However, the race was on, with the project almost completed. Bridgett continued living on any and all caffeine to stay awake, continuing to plunge forward to the completion of the project. This would be the "means to an end" with the government and their officials. Bridgett took advantage of the insomnia, as it allowed her to stay awake and think through many obstacles. One of which included the board of scientists that soon would approve or disapprove the projects conclusion. The lack of sleep also assisted in accomplishing ideas and thoughts regarding the safety of the program. The grand finale would be the confirmation that gives credence to the immortality for mankind. Bridgett understood the pros and cons of the project, as it justified the belief to ensure its safety. "Keeping it out of harm's way" would mean she would also have to market the project. As Bridgett's thoughts continued racing through with momentum and enthusiasm, an invasion of other mysterious circumstances took place. These adversaries of conflict were far worse than anything the government could produce. They were evil, bringing a turbulence of unexplained thoughts and events. And Hell followed!

Bridgett could easily sense the change in the room's atmosphere as it became thick and heavy, making it difficult to breath. These strange feelings were announcing something evil awakening deep within her soul. The dreams brought pain and agony to Bridgett as she descended into the pits of complete darkness and the abyss. There were screams emerging from everywhere within the darkness as they presented dark shadows that gave manifestations toward being inhuman forces. They tightly grasped her appendages stretching them apart introducing some type of torture device. As the dark shadowy figures came from everywhere within the blackness, their strength left no possible way of escaping. The environment of echoed screams were now recognized as her own. The sensation of unidentified torture items were being shoved into every orifice of her body causing severe sharp stabbing pains as well as a terrible burning sensations up inside the genital areas, with many repeated actions. Unable to close her legs, gave easy access to her vagina and anal areas that were thoroughly being mistreated. Bridgett's head was now being held in a vise grip that was being used to apply pressure to her head, while maintaining control of its placement. It turned into squeezing, slowly and then escalated. This was the same method used when a body is cremated-crushing the skull before entering the furnace. The tortures and screams continued with no time barrier attached. Bridgett couldn't release herself from the bondage but continued thinking of ways to escape. While being tortured. She would wait until the exact moment when her appendages were freed, then she would kick with every ounce of energy left at all the entities within the darkness with attempts to be freed. As the tortures continued, and no way to escape them, she prayed. Suddenly, as if God intervened her appendages were freed. She began the attack. Bridgett struck out and about in a frenzy through the darkness, with all efforts of hitting and kicking the entities living within the pitch-blackness. Hoping beyond all hopes that all the blows landed productively. Finally, the stretching stopped, and something new was added. It was as another unknown torture devise used as a whipping item and it tore through her skin with repeated strikes. Bridgett felt these lashings, as they seemed every whip took off layers of skin. The pelvic and breast areas took the worst beatings. Another torture device was added that brought thoughts of a red hot branding iron as it scorched into her groin, as she smelled the burning flesh. This tormenting device was symbolic of branding animals to be identified by their owner. The pungent odors of burning flesh and decay filled the darkness. Bridgett could only assimilate the possibility of being dead with decomposition taking place. The punishment which contained all the torture techniques became too incomprehensible for any human being. Since the brain was the last organ to shut down after death; she waited for all thoughts to cease. The hopes of escaping were gone, designated as an impossible feat. She was kidnapped by unknown black shadowed entities with only outlined appearances, and held captive against her will. Bridgett knew that whatever she had encountered was a master of torment and sinister evil attached. Everything would depend upon the fight to survive. That would be the only escape route to freedom. As the lashings commenced, Bridgett continued to fight, swinging her legs, hands, and feet, and would not stop until she was freed or dead; which ever came first. Then it seemed after that final thought and the huge struggle, she woke. Sitting dazed and confused Bridgett's belief was that the hounds of hell had taken hold of her, and would return to complete their dramatic performance. As she slowly moved from the bed to the shower every effort was riddled with excruciating pain. Every step was produced in slow motion. However, the warm water was very soothing, like a moment in heaven, to her battered body; which became the aftermath of the dreams. Bridgett began to wash the validation of the dream's tortures, as they were clearly evident. The worst afflictions were in the groin area, where she had felt the hot unknown device sticking up inside the vaginal and rectal areas. The bleeding and other fluid substances that flowed heavily from those areas gave indications of being repeatedly raped and by many. There were undeniable hot red raised areas with swelling and blisters and horrible open and disgusting sores. As Bridgett showered, the locations of the torture were evident everywhere upon her body. The hot water brought them to the surface, making them even more visible. She didn't want to think about the dreams and the tortures, and pushed them away as though they never existed. However, the mirror never lies, and the results from descending into Hell were validated, making forgetting an impossible task. Standing in front of the mirror, staring back at her was a reflection that once resembled that someone she knew. Bridgett's long dark black hair was left in a tangled mess entwined with dried clumps of vomit and blood. There also seemed to be some sort of wording scorched into the blackened area left over from clotted blood upon the pelvis. Approaching closer and closer to completely decipher the letters her hands automatically went over her mouth to muffle the screams of terror from the words that read: *"We will be with you always."*

Bridgett did not go to work that morning. It had been the first day she had missed in an entire year. The bleeding from her vaginal and anal areas were very intense. She remembered that ice would slow down bleeding and ice was also good for bruising. However, there was not enough ice to apply onto her entire body. Bridgett chose to stop the bleeding by compressing towels between her legs; while knowing if the bleeding didn't stop the hospital was next on the agenda. How would she explain the rape and the massive open wounds scattered all over her body? She would have to tough it out and continue nursing herself back to health. The migraines from childhood were returning. Bridgett was aware of how to get through them, and that was to lie down, lights out, with no noise, and applied those techniques. Resting at home from the aftermath would also bring pieces of the dreams to be analyzed. However, there were none. The only thoughts the dreams brought were the tortures. There was no memory that would establish a productive origin or meaning of their existence. The dreams never played out a scene, never showed any objects or clues, and knew it would be impossible to fight something that left no memory. Bridgett believed their real purpose was to cause vulnerability, helplessness, and leaving poor physical endurance. The essence of life was being depleted little by little each time the dreams attacked, leaving no self-control. She survived! No matter what the dreams brought, she was a survivor. Using her own strategies to stay alive. The dreams were ruthless being savage in their embracement of many different types of tortures, mentally as well as physically. The only thought embedded deep within Bridgett's mind was the certainty of them attempting to keep her and never letting her go. This would make the fight to awaken almost futile. It was quite hopeless trying to escape while preventing her to awake.

Their other purpose was to continue attacking until she begged to die, which would produce an enjoyment to them, she was sure. Bridgett believed that the strategic part of the dream would be to awaken as quickly as possible. There would be a continuous struggle to overcome these unknown forces. If unable, her fate was sealed. She would belong to them. The darkness with the outlined shadows; the inhuman entities lurking with anticipation of the tortures that waited to be inflicted. Bridgett associated them to resemble everything that was hideous and grotesque. The questions continued to pour forth. What are the dreams? What is their origin? What is their purpose? Bridgett tried to remember pieces from the dreams that might have been left behind. Some type of remembrance besides the torture. The episodes of these black voids left no trace elements to break through the realm of their existence, and there was one definitely created. It had left a message that could not be ignored. Bridgett felt that the dreams deliberately came forth with a connection to her subconscious as she awoke. The intruder that had entered Bridgett's body, the entity that came forth with power, strength, and vengeance, had a name. Yes, Bridgett recalled. It did have a name, but the migraines were blocking it, which made all thoughts cease. Her mind kept repeating *Remember its name! Remember its name!* Then the dreams allowed her to remember, but the damn migraines blocked it out. If they would subside she would be able to *remember. They gave her something, it was their name, yes she remembered this very clearly as the migraines vanished and her memory became alive as a piece of the puzzle to the dreams was freely offered. It came quickly and swiftly, entering at full speed and announced themselves so she would always remember them . . . **dark dreams!***

The new ideas for the project were very clearly intact. Bridgett was worried that the dark dreams might affect her competence to continue with the project's research, but they didn't. To her surprise, all thoughts during the brief amounts of sleep she obtained were actually beneficial in bringing the research to a productive and conclusive state. This gave Bridgett a sense of security in the knowledge that the project was ready for government inspection. It also added a new outlook that made her position over the project steadfast and well-armed when the government was ready for inspection. New characteristics were formed, conveying confidence and defiance. These new and different strengths emanated from somewhere within that brought her self-confidence to a much higher level. She had no fear of the government's denial of the project. In fact, there was a predicted outcome of positivity flowing continuously that made everything infallible. Bridgett's new discovery of these abilities produced convincing attitudes which were understood to be gifts that evolved from her profession and enhanced by her beliefs and morals. She was also convinced that there was a small sliver of light forming from within the abyss.

Tokens of approval yielded positive signs of the project becoming a reality. However, Bridgett remained in denial when relating the hatred of the dark dreams to the project's conclusion. They began throwing curve balls, offering a twist of fate; which linked an admission of guilt. Stemming from deep within there ignited a conveyance between the dark dreams and the project. Bridgett was afraid to even think or speak of them. Any and all connections between the two were hastily pushed away, shoved back somewhere behind the concrete wall where everyone was forbidden to enter. The random thoughts connecting any parts of the project and the dark dreams were abolished. Forcing herself to expel the thoughts of any connection though, left no reprieve. The realization at some point would emerge, and with it, there spawned a belief. Did she really believe the outcome of her project had anything to do with the dark dreams? The theories for science, biology, and genetics would open new pathways and endless avenues within the medical and science fields. Bridgett's scientific breakthrough would bring new developments and techniques for the future of mankind. As a believer of karma, she knew there would be a price to pay for these acquired gifts and strengths that were flowing freely. Bridgett was too involved in winning the Nobel Prize and was trying very hard not to mix anything that would remotely give any credibility to the dark dreams. She would never praise them for anything except the terror and fear they brought into her life. The price had no basis toward the conclusion of the project.

She was convinced that all dues were paid during the research process. Bridgett continued to disregard the karma factor by repeating over and over in her mind, *"I completed everything necessary to convince the government panel what was needed for the development of stem cell research and embryonic implantation. This was my achievement. These were my ideas, my dreams, and my research."* The ability for Bridgett to rationalize these convictions would justify the means. The focal points enabled the continuance of reaching the goal of the project. Yet to Bridgett's dismay, there was no exact truth to prove or disprove these thoughts.

Bridgett surveyed the lab and the scientists' multi research projects as they plummeted to the ground. She felt that many great science projects were being jeopardized daily because of the low budget and lack of knowledge offered to the government. As the labs were being closed left and right, she would jump, as if gun-shy every time a door slammed or a confrontation was taking place. There were many occasions when scientists would collide with government officials, almost initiating blows because of their indifferences. Bridgett felt very sad observing some of the best scientists with incredible research data enter the black hole. What a pity, Bridgett thought, shaking her head in dismay. All the dreams, which sprang forth beliefs, the time, and efforts placed into each and every research project that were once displayed in the gigantic quorum of labs. She truly understood the frustration emanating from their loss. Watching all the scientists she had worked with side by side for years, and knowing them quite personally as well, was devastating. These were professionals that were highly respected and admired within the field of science. These individuals were endowed with intelligence and greatness and were admired by many for their dreams, research abilities, along with their extreme enthusiasm, beliefs, and morals. She was now watching them slowly die one at a time, these believers of greatness, as their projects ended abruptly. Years of work laying in shreds upon the floor and discarded as unwanted trash. These projects could have had a significant and positive goal for humanity. Now they were bringing nothing to the table, not a mere morsel. Bridgett stood in the middle of it all as a cascade of tears, that could no longer be suppressed, streamed down her cheeks, as they became part of the madness. The anger kept building as the mayhem kept unraveling. Left alone, standing in the midst of total dismay and upheaval, time stood still. Now, the hatred toward the government took on a whole new meaning.

The first day of attending public school was the worst day of her life, Bridgett recalled. It was a very frustrating time as she tried understanding the way schools functioned, which brought about terrible anxiety and migraine headaches. This upheaval provoked actions that affected other students and teachers. The result was being classified as a loner and given the title of a nerd. She would be forever grateful to her mom for understanding the special gifts Bridgett was endowed with; along with acknowledging that this factor was a requirement for needing a special school. Mom handled it all very well, and never blinked as Bridgett was soon enrolled into a private school for gifted children. Then she became a member of MENSA, as encouraged by the other students. This was a place where they all came together and bonded. This private school introduced her to many different classmates who all seemed to have special gifts. She found it easy to mingle with them, and was fully accepted for who she was as they all blended together like peanut butter and jelly. The school introduced Bridgett to a whole new world of friends and, later, colleagues; they became inseparable and bonded for life. Bridgett often would overhear her family members say, "She has a brilliant mind and has become somewhat eccentric." She would always remember that one characteristic that stuck out more than the others when she heard someone say, "just a bird of a different color, that's for sure." She liked that expression and felt proud to hold that title. Bridgett's graduation from college as a scientist with many valuable credentials didn't surprise the family at all, especially dad. At graduation giving her a huge hug he whispered, "I guess your prepared to find the answers to all your questions now, aren't you." Bridgett learned later in life that being a scientist would, at best, be defined as a "different caliber" of profession. She compared them with that of late great poets, painters, writers, or composers. Some people even labeled them as eccentrics. She also expounded upon them as having a short fuse and a temperamental side. These characteristics were very familiar. Bridgett had a bad habit of trying to analyze everyone and everything. Although her best asset was attacking peoples mind. In this way, she knew them before they knew her. This gave her the edge on everyone without any suspicions. Bridgett became highly renowned for this technique. This pissed off a lot of her friends as they told her "Stop being a scientist when you're off duty. You don't know when to draw the line. Just have fun and quit trying to probe into everyone's mind and behaviors." But to everyone's dismay, Bridgett never took their advice. *It was the nature of the beast*, she thought, and felt a connection with all the personality traits involved with being a scientist.

There was just darkness, pitch-black darkness, showing no hint of light to distinguish images. The smell of rotting flesh was very distinguishable. She believed that *Once you smell the odor of decomposition, you don't forget the scent.* Recognition of that factor became present as it emerged from somewhere within the darkness. Bridgett gagged and knew she had to vomit but was unable to awaken from the dreams. As quickly as it came, a new smell arrived, and this one was that of freshly dug dirt. The heaviness of the dirt weighted down upon her entire body especially the chest, causing difficulty in breathing. Bridgett's nose and mouth were now filling with the dirt, and screaming made the dirt enter faster. Was she a rotting corpse? Bridgett's next thought was that she was being buried alive. The dirt continued as if equated to six feet. She began thrashing her body to avoid any further dirt on her face. The flight or fight syndrome took over, as despair took place, trying to awaken from the dark dreams. The use of her hands and arms became instrumental in pushing upward through the dirt, up and up, attempting to dig through to the top, hoping someone above would see them, and come to the rescue. Picturing that aspect in her mind gave hope and the confidence that someone would surely see what was happening. Although another mental image took place also and it wasn't very appealing, as that of a zombie coming out of its grave. Bridgett could feel a small amount of air thus making an assumption that freedom was close. As all energy was drained, and exhaustion took over completely the lifeless body that once was alive with all the passions that life had to offer was failing to survive. The dark dreams arrived causing a surrender to the darkness. When the feeling of being suffocated slowly subsided; she knew the dark dreams were releasing their vengeance, and would awaken. The aftermath of the dark dreams was noted everywhere as usual leaving huge amounts of vomit from the bed to the floor, over to the walls, and reaching as high as the ceiling. Bridgett jumped from the bed and scurried over to a mirror to assess the damage made by the dark dreams. There in the mirror staring back was someone almost unrecognizable. The copious amounts of vomit and dried blood brought the appearance of a native preparing for war. Her hair was so matted it resembled dread locks. Looking back at the room, all the bed linens were ripped into shreds, irreparable, as well as all clothing as if a wild animal with sharp teeth or fangs and long sharp claws had been let loose while she was sleeping. Bridgett's first prediction was right. They were the Hounds of Hell being unleashed to finish devouring their tantalizing meal. Everything was viewed in disbelief while standing in the middle of the chaos. There continued to be gut-wrenching nausea, creating dry heaves while being subjected to the foul odors lingering in the air. Still, the dark dreams did not produce anything of value or interest. There was nothing all she could remember was the torment endure.

Bridgett began to believe that maybe she was involved in some type of paranormal experience as she continued to stare at the room in disbelief. Then a thought emerged as a question that made complete sense. *How could I have produced this much vomit by myself?* This logical explanation made it perfectly clear that she was not the only participant. Bridgett wondered if there would be any serenity in the future as the dark dreams continued to stay and make their home somewhere within her mind. As a scientist, Bridgett needed to formulate a hypothesis as to their origin. However, the dreams left nothing, no remnants that could be pulled forward to analyze except their name. Other reminders were the words embedded deeply upon her groin area, along with bitter sweet mental and physical anguish. That was all that could be taken or remembered. The aftermath of the dark dreams were Bridgett's greatest fears, because without them, the reality was she was captured. Even though the dark dreams left terrible displays of destruction, the surreal part was she still survived.

There was an unusual experience that occurred one day at the lab that would change her life forever. It involved a stranger assumed to be one of the government's officials who approached her resembling an official but was not wearing his badge. She knew that every representative had to wear their credentials within the confines of the lab or pay heavy consequences for breaking government rules. There were no exceptions to the rules under any circumstance. These badges gave credence that they were representatives employed by the government. This pretense of an official approached her and whispered, "Bridgett, I need to tell you something off record." She kept on working but listened intently as he spoke. "I know individuals within the field of science that belong to a powerful group that call themselves the Secret Society." Then he quickly pushed something into her hand. She stopped to look at what he had given her, and it appeared to be a business card with just a phone number and nothing more. He continued whispering, "If you call that number, leave a message with your concerns regarding your project. They will return your message no later than twenty-four hours." This stranger also told Bridgett that the people involved were very powerful and unknown by the government. Then as quickly and quietly as he came, the unknown stranger left. Bridgett nervously put the card in the pocket of her lab coat and continued to work as if nothing happened. Her mind was left spinning about this stranger who posed as an official and the racing thoughts began turning into paranoia. She began thinking there was a conspiracy taking place and started looking for the cameras and where they might be hidden. *Was this a setup?* She kept asking herself. *Would they be sneaky and desperate enough to get something on me so they could take the project?* Bridgett always seemed to be on a seesaw with the government, along with the dark dreams. Heading home that night, thoughts regarding the day's events led to a decision that was urgent. Bridgett would give it a night to formulate an opinion.

Much to Bridgett's surprise, the decision came easily the next day at work. There were twenty-five officials the day the stranger approached and placed the card into her hand. Bridgett made it a point to count every official upon entry into the lab, because it gave her an extra card to play if there were any surprise attacks. While searching every face that day, his was not among them. He knew her name, but she never knew his. Bridgett couldn't stand the suspense any longer. When she was about to leave work, there was an official exiting the lab also. Bridgett put on her radiant smile with a flirt attached to it and asked very innocently, "I saw twenty-five of your people here today." The official responded, "No, there are always twenty-four of us assigned to this lab." "Oh"—Bridgett displayed a shy look—"I must have counted wrong, sorry." Then she gave another cute smile and disappeared quickly into the parking area. Upon reaching the car, she heaved a sigh of relief. Feeling that everything was safe, and there were no suspicions attached to the question that would alert the government grapevine. Everything seemed right with the world.

The white business card was on the table beckoning her to make that call. Bridgett began pacing back and forth, thinking about the stranger and the business card. She no longer believed he was part of the government, due to the accurate count of officials. Bridgett found the situation so intense that it actually became exciting. She knew there had to be a reason for his forward approach. He was offering information about a Secret Society within the government. *Why me?* Bridgett asked herself. *Maybe*, she thought, *just maybe this was for real, and there was a high regard for my project.* Bridgett wanted to have a positive attitude, as negativity took up a larger portion of her life. If there were such powerful people who believed in the project's goal, perhaps they did realize the credibility attached to it as well. The stranger was just a messenger and nothing more was concluded. However, the lab was off-limits to everyone except the officials scientist. The project was top secret, and a military police soldier was located at each exit, armed with weapons and not friendly. They were not playing, as exhibited by the security of Area 51. *What balls! Amazing!* She thought of the stranger. Allowing herself to digest this unusual event and the complexity involved, Bridgett decided to take the chance and dialed the number. She left her name and number bluntly asking if they could rid the government from being glued to her ass. Bridgett also asked if there was any way to get more funding for the project.

The next day at the lab, Bridgett watched the clock, as the day moved in slow motion. Within twenty-four hours, as told by the stranger, it happened. Bridgett was amazed at what took place the next day at the lab. There were no further issues with government officials. Instead the officials were offering praises, and announcing her compliance with the project's time frame. The lead official said, "Take all the time you need, Ms. Montgomery. We completely understand the importance of this project." Ms. Montgomery! None of the officials ever addressed her that properly. She was just called "Bridgett." Everything that came forward from within yelled loudly, "It worked! There was a Secret Society, and they approved all my requests." It was as though magic fairy dust was blown over the entire lab, making all the dark clouds disappear and bringing in sunshine and lollipops. Bridgett didn't care how everything was accomplished; she just enjoyed the bliss as everything about the project started falling into place. The only problem Bridgett could surmise was that the Secret Society wanted something in return, and sooner or later it would emerge. She understood that somehow they were engaged in her life; the how or why would have to be set aside for now. Bridgett observed the longevity of the officials as they surrounded all the other projects within the lab. The stress was obvious as viewed upon her colleagues' faces, also indicative of gritting their teeth when answering all questions. Bridgett laughed with the thoughts that remained about them, *still in comparison to roaches.* Bridgett was now completely left alone and free to engage the project with more positive research to ensure the safety of its ability when used on human beings instead of other living creatures.

When Bridgett returned home the blinking message light was on the phone. A mechanical voice was used, giving instructions. They were emphasizing the government branches that would be beneficial for the project. Bridgett kept notes on everything that was being communicated, as she deemed their importance and would need to follow every detail. There were no room for errors! Bridgett's conclusion was that the fate of the project was now in the hands of the Secret Society. There was apprehension upon entering the lab the next morning with worries about the grapevine. As the day coursed on, nothing was ever said. Knowing the government and how they would jump right on it immediately; gave meaning that she was clear for takeoff. The project remained safe and alive.

Bridgett became obsessed with checking her messages, as this action was very new. Then again, everything seemed to be a new experience. The light of the answering machine was blinking upon entering her home. Bridgett was anxious to hear what would transpire next. The message that was received totally blew her mind. It began with, "We know you're worried that someone will find out about us, our involvement. We genuinely understand the greatest fear you're facing, and it is the loss of your project that was created with such loving care, while proving its safety and concealing dangerous research. You can rest assured that nothing will ever come between you and the safety of the project even after the final development. We believe in your work and will follow when it comes to a conclusive state, before the board of scientists. The knowledge of stem cell implantation and embryonic implantation will produce a new era in science and medical technology. You must believe that we cannot express the importance of this achievement. There is nothing to fear from us. Everything will turn out in your favor. The only thing we ask in return is that you provide this society with its continued privacy without any exploitation. We are the keepers of the cards and only offer them to the chosen few." After the message, Bridgett had to lie down as breathing became erratic and she became light-headed. Her belief was that she had fallen into a time warp, a parallel universe. She felt that somehow they had tapped into her brain; where all the fears and secrets were hidden. Behind the block wall. All she could feel was positive energy toward the project's resolution. The "Atlas" projection occurred, as though the weight of the world was lifted off her shoulders. "Atlas" in Greek Mythology was the son of Zeus and punished his son for coercing with the enemy, and condemned him to stand at the western edge of the earth and hold up the world upon his shoulders, to prevent the two (the sky and the earth) from resuming their embrace. Also the sword of Damocles that was hanging over her head with impending doom left as well. The "Sword of Damocles" in Greek Mythology was a poor man and was jealous of someone that was rich. One day the Ruler of the court told Damocles to take everything he owned, gold, jewels, food, fine clothing and included his position as the court ruler. So Damocles took everything and was enjoying it all until one day sitting at a table ladened with all types of foods that only the ruler of the court could obtain, he looked above his head and saw a huge sword with just a hair holding it up. Damocles asked the ruler if he could take it down, and the ruler said no it comes with everything you wanted. Damocles was very upset with the ruler and said, "How do I know when it will fall, and kill me?" The court ruler stated, "You don't, but as a court ruler you can not change it. Damocles thought to himself, "I don't want to worry each day that it may be the day the sword falls, that is too big of a worry." With that terrible dilemma in mind, he quickly gave everything back to the court ruler, and went back to just being an average person with only an adequate amount of riches. Until you walk in someone else's shoes you should not be envious of what they have.

Bridgett was immensely excited and very proud that she was highly respected with many powerful followers. Her intuitiveness gave an insight into these influential individuals, believing the Secret Society had been observing since the start of the research. Bridgett felt for the first time in years that there was a family within the realm of science who were trusting and protective; while keeping the safety of the project from all enemies within the government. She was very firm about those who took her niceness for weakness; however, that was not occurring with the Secret Society. They knew the strong characteristics she carried which would not be easily defeated. Maybe that's why Bridgett became one of the chosen.

The time was now to push forward in one gigantic leap for a conclusive and positive ending. Bridgett could wait no longer, and worked fiercely, by passing all days and times staying highly focused on one thing—the completion of the project. The confidence she was feeling rid all nervousness performing under the spotlight in front of the Board of Scientists. The conclusion had to be smooth, well-planned, and given step-by-step research, with all stages encompassing the entire project. The expectations of the board were harsh and adhered to strict guidelines, which had to be followed in sequential order. The hypotheses that led to her research had to be thoroughly understood, because the demand for stem cell and embryonic implantations would break all barriers of science and technology. This discovery would trigger the rebirth of a new era, which Bridgett would be considered the creator.

Standing poised with appropriate attire for the occasion, in front of the podium using correct and professional eye contact. The spotlight blinded her from seeing the faces of the audience. She knew the board was present, and at that precise moment in time was all that really mattered. The audience contained people from every branch of the government along with science colleagues. The board was prepared with its highly renowned and accredited scientists. She gave them all the information that was acquired from her three years of devotion with the project discussing the theory of stem cell implantation and embryonic implantations. The audience remained in silence. No exchange of air was heard from anyone. A new era of dynamics was needed for the preservation of mankind. The project was proven to extend the longevity of human life; while preventing any further suffering and pain for humanity. The conclusion kept flowing and the knowledge of the entire program was emptied before the board. Memories of the project flowed gracefully and easily as that of a child's ABC's. The validated facts were not read but spoken from some place embedded within Bridgett's mind and soul. She projected an awareness that took on a role of mother to baby, as depicted by the Board as well. She expressed the nurturing that took place from a single embryo in a petri dish to a complete and thorough developed conclusive state. There were no questions posed to her that could not be answered without hesitation, Bridgett's confidence remained on high peak. On that day, before the board, nothing formulated from within her mind or filtered through time to alter the project's potential. Bridgett's presentation to the board was flawless.

Bridgett Montgomery had won the Nobel Prize for molecular biology. She was honored with bringing new and safe technology into the world and was noted in *News World Magazine*. The project created was now alive and waiting for those who believed in its potential. The top-secret project no longer had to be hidden, and the name of Bridgett Montgomery, PhD, would never be forgotten and would soon be a household name. The doors were wide open for anyone to use for the purpose of what had been created.

Bridgett knew from the beginning that she had the dark horse. Realizing it was a winner from the start; when it was but a single living cell. In a state of disbelief from the results all Bridgett could say was, "It was amazing!" Her technological achievements were accomplished not out of selfishness for money or fame. This characteristic was what made Bridgett a different breed of scientist. It was the genuine love of humanity. The miracle baby was now recognized for its intent and purpose. She would always insist that the project would continue as part of her life. The bonding that both of them shared would remain inseparable. Bridgett's approach to the government was the understanding that she would always stay in control of its existence-permission granted. The validity of the research had been thoroughly acknowledged as being safe. Her achievement that was based on genetic engineering was ascertained by the government as being the biggest invention since the nuclear bomb. *Terrifying*! was Bridgett's thought to that comparison. This acknowledgment by the government was subjective at best; which literally made her sick to her stomach. She was left in shock and numbness. *Hum*, Bridgett thought, *save lives or destroy them?* Tough decision to be sure, as she thought of their idiotic statement. There were no words in her vocabulary to describe the comparison they presented, and wasn't going to waste any time trying to understand the government's dysfunctions. Her entire life had always involved people; which most scientists schied away from in the public eye. Bridgett welcomed them. She wanted to do more in the marketing aspect of the project, while someone special needed to run the labs. The dilemma of hiring for the labs and marketing were too big for her to tackle. There had to be someone trustworthy enough to do these tasks. Beyond any doubt all guidelines, per government protocol, had to be to adhered for the safety of the client. Bridgett beat herself up, trying to think of who to hire that was trustworthy without question. Her burned-out brain, still hanging on to its sanity, finally quenched its search, as the lightning bolt struck. The rainbows soon appeared with its beautiful hues of color as Tom's name entered her mind. *Why didn't I think of him? My god*, Bridgett thought, *Tom had supported my ideas from the start and knew everything about the project.* She would beg his forgiveness for having no time allotment to share, and he would give in easily as being highly pleased that he was chosen. This would give him ammo to brag about her hidden love for him.

They had known each other since childhood, and both were involved with MENSA. Tom Thorp was a personal friend and had dated sporadically throughout the years. The truth be told, Tom had a deeper attraction for Bridgett than she had for him. He would boast on many occasions to their friends and colleagues, "Someday she'll be all mine, and I am willing to wait. I don't care how long it takes for her to make a commitment." This made Bridgett very uncomfortable, although she adored his personality and knew his faithfulness held strong convictions. Luckily he never became a fatal attraction. Tom never treaded in Bridgett's territory and was very respectful, waiting patiently for her calls. He never showed aggression, and she enjoyed his personality. Tom thoroughly understood her strong characteristics as well. He knew without a doubt that pushing Bridgett in any direction would be his loss. Time was spent mastering his skills and abilities for her to play in his court without realizing how she got there. It would be a master plan for sure Tom thought. Bridgett always stayed within the circle of friends, as they kept her updated about many things, especially Tom. Not that she wanted it; they just assumed it was necessary due to the bonds they all shared.

Bridgett's thoughts of Tom were realizing how pathetic his needs for her remained so deeply rooted. Their colleagues and friends understood Bridgett's free spirit and noncommittal attitude, as did Tom. Therefore, no one felt any sympathy, only blame for his personal desires for her. It was very obvious that Tom didn't care who knew about his eternal love for Bridgett. They did value his opinions very highly, as they always called upon him in times of a crisis. Tom was very good in dealing with stressful issues. He was laid back and a very good listener; which made everyone very comfortable. People opened up freely with communication as did she. Bridgett knew that Tom's IQ was off the charts, and was attracted to that particular characteristic. He carried around strong convictions about their future together. Bridgett took it all in stride and never freaked out over Tom's obsession for her. She never had any thoughts of having a relationship with him. She needed his friendship, and he was a willing participant; it was that simple. There were many uncountable times when she asked for his advice knowing without any hesitations that his personal feelings never interfered. Tom believed in the project and at times pushed her toward the project's goal; when all Bridgett wanted to do was throw in the towel. They both shared the same beliefs and morals that were associated with the project. At this juncture, he was the only person who made any sense in her life. His credentials were valid, as he had a BS in biochemistry and microbiology with a PhD in microbiology and immunology. This made him a very valuable player. Every time Bridgett marketed a lab to be opened, it was followed with a rapid response to its needs and supplies; which included employees. Bridgett needed Tom's assistance to open labs and to hire professional staff qualified in different areas. She knew from every aspect of her soul that he was the only person that could be trusted. Bridgett wanted him to know that the project came before any social life, and he agreed. Yet on the last day she spent with him, he looked straight into her eyes and said, "You will be mine in the end. You know that, right?" Bridgett never responded back with any comments; the wall was being built of cement and blocks.

The mail was retrieved from her apartment lobby, and as she flipped through it, Bridgett's eyes stopped immediately as they focused on a blank envelope addressed to Ms. Bridgett Montgomery. *Now what?* Were Bridgett's initial thoughts. *It's never ending!* She was beyond the point of acceptance with everything and everyone. However, the inquisitive part of her profession took over as she savagely tore open the envelope. There inside was a gold plastic card with black bold numbers. Bridgett could hear the pounding of her heart as the thump, thump,thump blocked out all other noises. The light-headed episode returned as the dizziness felt like the world was tilted off its axis. The trembling and shaking of her entire body brought back old scars and memories that reopened when she read the note attached: "This carte blanche is specifically for you, Bridgett. We will always be here if you need us." Although this mirrored the past of their first encounter she knew the card could be beneficial, and kept it. The Secret Society always seemed to forecast future events of problems, but at this time, there appeared to be none. This left Bridgett with a taste of a bad omen that was probably about to transpire. She exclaimed loudly while waiting for the elevator, "For heaven's sake, what else is going to happen?" Bridgett wondered if there would ever be solace, and the need for the card would not be necessary. Sitting in bewilderment within her safety zone, thoughts returned concerning the Secret Society that inspired a new conclusion. *Why would I want to get away from them?* She had accepted them as family and was never abandoned; that allowed Bridgett some peace and tranquility. Maybe it was just the creepy feeling of them knowing her every moment. There was probability that they could even have knowledge about the dark dreams. Bridgett was now overtaken by negativity as she bore deeper into waves of foreboding thoughts. *Why are they monitoring me so closely? Are they spying to see if I broke our bond by exploiting their existence? Were they expecting me to sell out to the enemy? Were they thinking that I would become careless, allowing the government's specifications for embryonic implantations to fall through the cracks and not adhere to the guidelines issued to the program?*

The new procedure for embryonic implantation and stem cell implantation had finally reached its destination to the public. In this acknowledgment, Bridgett's notoriety grew as well. This overwhelming stardom gained momentum at an alarming rate that took on a whole new level of existence. Outwardly, she appeared to the public as being a glamorous and sexy female scientist, who just happened to be single. This exploitation made their way through many national magazines with pictures of her as well. Bridgett did not enjoy this particular type of publicity. It gave the impression that to be successful, beauty was involved. These types of remarks made it difficult to defend the feminine movements based on intellect and not necessarily beauty. Bridgett's achievements were slowly being replaced by what she referred to as garbage news. She enjoyed reading about herself in science magazines, and taking center stage on TV talk shows. These interviews triggered a source of intellect based upon many issues that were enjoyable as explaining the project. Bridgett Montgomery was now known Internationally in the spotlight. This produced opportunities to share the stages of the stem cell implantations and embryonic procedures that were important and crucial. It also enabled her to answer all questions that were interpreted as fears. The acknowledgment of endeavors in both fields of medicine and science technology combined, brought both support and criticism. Bridgett was now able to reach those individuals who had waited patiently for the program to be legally opened. These were the people that shared the same interests morally and ethically in sustaining the longevity of life by thoroughly understanding the stem cell procedures, and embryonic implantations. Bridgett was very eager to share all knowledge behind the project, and more than eager to accept embryo donations. It was settled. Tom would run the labs that were already opened, she would market the project with good intentions to save mankind.

Bridgett could feel them creeping little by little, hiding in the shadows wait impatiently for their arrival. Then from the darkness, they hit as anticipated with a TKO. A giant wave of gut-wrenching pain forced Bridgett to double over, and then she was held down in that position by unknown forces as the acid poured into her throat burning terribly. The taste of the acid remained and could not rid herself of it. The nausea and vomiting had become part of the dark dreams, as well as many foul odors she was forced to endure. Their inhuman punishment left no short degree of mercy or reprieve. The pitch-black abyss was now everywhere and brought with it the fear of being hunted; waiting for the apparitions to take their places within the darkness. The entities once again took hold with their relentless strength pulling in different directions; while deep growling noises was heard. The Hounds of Hell were Bridgett's thoughts as they seemed to be fighting over the amounts of her flesh claimed as food substances. The sharpest pains imaginable took place as reflected by the sharpness of teeth, or fangs with the feeling of long and sharp claws penetrating into her flesh. Bridgett thought she was giving birth, and the fetus was being cut out of her womb by a sharp instrument, without anesthetic. Sounds entered into the silence of the black void of gnawing, chewing, and slurping; while the growling commenced. They were gluttons, devouring and consuming as much food as possible; while fighting over their territory. It was very clear from the pain and sounds induced that she was being eaten alive. Bridgett prayed to pass out from shock or just die. The icy water thrown upon her purposely kept her awake to feel more pain. The slurping noises were now recognized as fluids being sucked from her body. She was glad that the darkness covered the grotesque figures as they feasted upon their human catch. It would surely provide a horrific display for anyone if they were visible. The agonizing reality of being eaten alive left no human comprehension of what was actually taking place. Screams continued to echo from every direction, but none sounded human. Hers was the only human voice ever present in the dark dreams. Listening to the animal noises just convinced the imaginable as to their inhuman ugliness that would be too terrifying to view. It seemed each time the dark dreams took over, the element of time seemed to be in their favor. The length of which dragged on to eternity. Bridgett struggled to free herself by awakening, and they continued to suppress this ability. They seemed to be sitting upon her chest, and they were very heavy causing great difficulty to breathe; while trying to free herself. Now out of the blackness the dark shadows grasped her around the neck in a choking manner. Weakening from the battles of struggling, Bridgett laid still, as they took whatever was wanted. Suddenly, from somewhere deep within the blackness, came a faint voice. It then excelled to a voice which became louder and louder as it drowned out the darkness with it's cries, "God, help me, please!" Recognizing the voice emanating from the depth of her soul Bridgett's prayers were answered, and was immediately released to waken.

When looking at the aftermath of the dark dreams, Bridgett would assess herself by scanning her body swiftly, as if a victim in shock. There were bruises, but the flesh was still intact. This was nonsense, Bridgett thought as she was aware of the tortures endured within the dark dreams, and just finding bruises? The neck region, however, did provide some evidence of what had occurred as there appeared to be bruises surrounding the throat area. Some bleeding trickled from her nose but stopped quickly. Bridgett's hair and face had dried vomit. This aftermath, as she called them, were quite common. However, she had to admit that the vomiting episodes were lessening. This time, it was only matted hair that continued to resemble dreadlocks, and the vomit was only displayed upon the walls, the ceiling, and the bed. The bedroom looked like usual after the dark dreams—displays of total chaos resembling the remnants of a slumber party. Pillows with stuffing flying everywhere, blankets thrown all over the room, with the sheets and clothes ripped to shreds from the Hounds of Hell, just the usual. As the days became nights and the nights became days, she felt that her entire life was becoming an unbearable prodigy of disaster. Bridgett's proclaimed characteristics as being very strong and defiant were lessening. Having a military upbringing was strict and harsh with no crying allowed. She was told to "have a strong backbone," that "crying showed signs of weakness," and that "you can do anything a man can do if you put your mind to it." "If you want something done, do it yourself." Bridgett wanted the respect of her family so all weaknesses remained hidden. Now with the onslaught of tears gushing out, she knew this weak display would have to cease. The significance of crying would only prove to the dark dreams that they were winning and that weakness was taking over. Bridgett was very fearful that perhaps they were getting stronger each time they arrived. These thoughts would need to be hidden, as their conclusion would mean she was pathetic. The word pity derived from that word was unknown leaving Bridgett waiting patiently for the dark dreams to finish devouring her remains.

Slowly drifting off into sleep and whatever it brought, she woke with a new concept. It was a good time to use that "carte blanche" and make a phone call. Unbeknownst to her at that time, this wasn't going to be an easy task. Bridgett found that she had to force her thoughts toward the direction of the phone and the momentum used to get there became a huge fighting event. While inching herself closer and closer toward the phone, Bridgett noticed that all the energy which had once flowed freely was now almost depleted. Ambulating to the phone had now turned into a procedure of using baby steps inching closer and closer to the destination. Everything seemed to take place in slow motion, as if gravity was causing the inability to get anywhere quickly. She was very aware that the dark dreams had everything to do with this problematic situation with the inability to reach the phone. However, Bridgett continued talking herself into every move by loud audible tones, which became a screaming match with the forces holding her back from reaching the damn phone. She kept repeating to herself, *"Get to the phone, Bridgett. Don't let them win. You can do this, it's easy. Just run and pick up the phone. They can't stop you once you get there. Then she heard herself asking, Where was the Bridgett that was invincible, who excelled above her entire class in science? Who and what did I become?*

The steps leading to the phone made her legs feel as though she just ran a marathon, with her knees buckling under the strain. She tried grabbing something for stability to prevent falling, as her legs and feet seemed to get heavier and heavier with every step. She finally grabbed a chair and used it as a walker. *What was this monster that I am keeping inside? This entity that calls itself dark dreams . . . what was its purpose?* Bridgett knew if she could analyze its origin or understand its thought processes, therein would lay the answer. It could think for itself, name itself, destroy, and take over her own thought processes. Bridgett soon noticed that her own thoughts were now expressing fears of the dark dreams. She tried not to believe that this was factual as their ability would possibly launch a whole new approach. Bridgett sensed its role connected with her life and hated to admit what was revealed. The revelation was that whatever involved the dark dreams was attempting to take over the life of Bridgett Montgomery. The only hope was in placing all the pieces of the puzzle together to find out what they really wanted. Catching it off guard was the only way it could be defeated. Bridgett didn't want to think about it anymore. She stashed it away as deep as possible within the soul for protection, not allowing the dark dreams to know her strategy. Bridgett placed the white card high in the air, announcing while yelling loudly within the apartment, "Watch this! I'll find a way to get rid of you, watch and see." Yet underneath that cloak of Bridgett's strong personality traits laid deadly evidence that encased fear. This frightening thought was that their characteristics were becoming much stronger; posing the greatest threat of finding the relation between the dark dreams and her life. *This is what I needed*, Bridgett thought. *I should have done this long ago.* Finally making the journey across the ocean, through the desert, and up the steepest mountain range, she finally reached the phone and dialed the number.

The Secret Society sensed something was terribly wrong with Bridgett. They tried not to discuss anything with her, but arbitrarily decided to make their physicians available as part of Bridgett's life. Their purpose would be to stop a walking time bomb. When Bridgett finally received a message from the Secret Society the call became a blessing in disguise stating they would help with her well-being. Staying always truthful and being very successful with all their endeavors never faltering on any of their promises; Bridgett trusted them without question. This removed all the doubts and questions connected with their falsehood, adding, *I trust them with my life*. Bridgett never tolerated lies. The words deceit and lies disgusted her worse than the dark dreams. "The truth will set you free" was the exact motto. The Secret Society and her remained tightly bonded. They sent a list of physicians with scheduled appointments.

The Secret Society came straight out with, "Only use the physicians on our list." Bridgett knew that they were not playing. All that was needed now was her appearance, but she was very bad at keeping appointments and conjured up many reasons for not going to them. Now there was a promise and commitment made stating, "I will continue with all physician appointments and referrals, because I need to get better." Bridgett knew that once they were called it was taken very seriously. If they were disappointed in any way you would no longer belong in their circle. They were the only people she could turn to for support without losing it mentally. Therefore, the instructions would be followed without any deliberation. Bridgett knew for damn sure that her name would not be recorded on their list of ungrateful people. Even though there was a slight danger involved with the Secret Society, Bridgett continued dancing to their rhythmic tunes without missing any steps.

The first physician, as she recalled, Dr. Creed, MD, reminded her of the rural physician who everyone came to trust and love. Just a small town doctor, yet he knew everyone's social dynamics and were known for keeping secrets. This made visits with Dr. Creed very comfortable; feeling that she had known him forever. After all assessments, Bridgett was assured that there was no significant health issues present at that time. The follow-up appointment with blood work, CAT scans, and x-rays were thorough as any good physician would recommend. The prognosis was possible exhaustion. Dr. Creed then placed a referral to a Dr. Martin Levy, who just happened to be on the physician's list from the circle of friends. Bridgett was now feeling a little more confident that perhaps there were no health issues. Realizing that it wasn't so bad keeping appointments, getting a second opinion would be a quick performance. Dr. Creed raved about Dr. Levy, and his ability to diagnose and cure patients that left other physicians mystified. Dr. Creed stated it was a gift he was given. Then he spoke further explaining about Dr. Levy's new high technical devices assisting him with many unsolved medical issues. He further stated with confidence and sincerity in his voice that Dr. Levy could possibly find a medical issue that was overlooked. Leaving Dr. Creed's office that day gave Bridgett a feeling of success along with feeling proud of herself for keeping the appointment. Finding out that a prognosis of exhaustion was hardly cancer was an added bonus. These good thoughts produced a speedy but happy walking pace. She scurried to the nearest store, ransacking the shelves, trying to find the most potent product over the counter to avoid sleeping. Preventing sleep that evening was very essential. It was important for her to look and feel her best so there would be no suspicions related to health issues. Bridget wanted to think just as Dr. Creed had interpreted—nothing but exhaustion. If the dark dreams entered her sleep and attacked, she would look and be a wreck at the appointment. Bridgett was now obsessed with never sleeping and avoided it as much as someone avoided the dentist by associating dental work with pain and discomfort. It was this form of anxiety confronting all thought processes trying to escape physicians who might find her secrets of the dark dreams. This would destroy all knowledge of their origin and purpose. After everything the dark dreams had stolen from her life including their aftermath of savage rapes, Bridgett took it very personally. She was obsessed with destroying them herself.

Dr. Martin Levy, referred to by Dr. Creed, was next on the physician's list given by Secret Society. Upon receiving a clean bill of health, she did not feel that her well-being was in jeopardy. Bridgett was aware that Secret Society wanted her assessed by all the physicians on their list. She did not want to disappoint them. Dr. Levy was a young doctor in his thirties, and was already head of the Cardiology Department at one of the largest hospitals in New York. Bridgett knew his background from reading science and medical news magazines and was aware that he assisted with the first organ donor program. Dr. Levy was also accredited as being the best heart physician, named by *Top Doc Magazine*. Many physicians that were very well-known became part of his physician's team. On her first appointment, Bridgett found Dr. Levy to be down to earth, very vocal, and straightforward, with a no-nonsense attitude. She was attracted to that type of personality. He displayed no power plays and no signs of conceitedness. Bridgett was very surprised that he made no attempts toward the God complex. While in his office each time Dr. Levy turned away, Bridgett's eyes began to scan the different certifications displayed as hers. Kept in beautiful frames signifying their importance. While he continued to review her chart, Bridgett kept investigating his office. Her eyes stopped scanning when she came upon one specific certification that jumped out more than any of the others. Dr. Levy was also a member of MENSA. She didn't want him to know that they probably shared many secrets; while continuing to be mesmerized. *How intriguing*, she thought. *He actually listens to his patients with a superb and professional bedside manner.* Dr. Levy had personal clients only, all others were seen by his colleagues. Bridgett understood the involvement with powerful positions and the spotlight that followed. The constant vigilance of outsiders provided clear and precise reminders of who they were within the public eye. Bridgett's understanding of that particular kind of life and all it entailed; she couldn't believe that he still made her a private patient. Now they just added more on both their plates. Bridgett voiced her apologies stating, "I came with no references, some friends of mine said they knew you well, and you wouldn't be upset as to the intrusion." Dr. Levy laughed, "I wish all my intrusions were like you." Then made light of the comment by stating, "There's no references requested on the very first consultation." However; she overheard him tell his nurse, making it very clear, "No one is allowed to speak with Ms Montgomery at all." Bridgett made a silent bet that no one dared to question his orders. The Secret Society was behind him, and believed this to be true while watching him walk on egg shells. She could see he was trying his best to gather any and all information that she was willing to offer. Bridgett often wondered if Dr. Levy was also afraid of their wrath.

Bridgett became a repetitive patient of Dr. Levy's while keeping her part of the bargain by attending all appointments as promised. The cardiac tests were very lengthy. But, she didn't mind any time spent in Dr. Levy's presence; because now there began an attachment to his ways, eg. character and personality. She wanted even more time to spend, as much time as possible, and suspected he felt the same way. This time allowance with Dr. Levy brought many thoughts of being single; which led toward a different direction of wanting someone close in her life. Bridgett believed more than ever that she was ready to have that love and warmth that came with a relationship, and would be willing to take a stab at it with Dr. Levy. He added something new and different into her life; which produced close feelings for him that went further than just physician and patient. Bridgett still knew that no matter how close they became; she would never divulge the secret of the dark dreams. There had to be an open and honest trust accomplished to convince a bonded loyalty. Bridget's persistent thoughts interfered always, as a reminder of, "*I'm not ready*," upon every visit. God, how she wanted to just burst out with the horrible secrets hidden behind the wall. The dark dreams that were tearing her life apart bit by bit inviting nothing but health issues, and possibly death. The worst part that the dark dreams brought was the disassociation from society, leading to loneliness and despair. Going thorough those scenarios still prompted Bridgett not to disclose them. Martin, as Bridgett started calling him spent a lot of off-duty time together. There were many walks hand in hand, having lunch's together and him preparing dinners at his white magnificent mansion. There were countless hours shared in each other's arms, and intelligent conversations that sparked many flames that began to kindle into fires. They had became very socially connected and pictures were taken of them together and exploited in the countless gossip magazines sold at stores. They both enjoyed plays and operas which sometimes took them out of state and away from her life as a scientist and his as a physician. They both enjoyed the same sports and took many out of state trips to see them. Their physical relationship was the biggest hurdle. Due to the rapes during the dark dreams, Bridgett wouldn't allow any sexual contact to take place. This, of course, raised many issues for Martin, but were quickly sloughed off by Bridgett. He didn't push the issue; however she knew he wanted answers. Even though none came forward, he continued with their relationship pulling Bridgett into his off-duty job volunteering at a hospital in the Bronx. He mirrored her own personality traits in so many ways. Their personal encounters became very intense and started to spill into her life.

These deep feelings he communicated almost reeled her in with a very appealing lure; but the deep-seated thoughts would not allow it to happen. *No, No, don't tell him!* The cement wall kept getting taller and taller while maintaining the strength and safety with restrictions on trespassers. Bridgett noticed that sometimes when their eyes met, it was if they both knew the answer to each other's happiness. This intense eye contact was very strange. The point of Martin being in tune with her thoughts, characteristics, and personality gave him an A plus; along with the warmth and confidentiality concerning his personal life about her. He was waiting for her to do the same, she suspected; however Bridgett told very little about her life. The worst aspect of their friendship, as it once was called, left their physical contact with just kissing and nothing more. It appeared that the more she became involved in his personal life, the more invincible the block wall stood.

Dr. Levy was desperately seeking an answer, Bridgett suspected. She began to wonder if it was about the dark dreams. *Maybe that was what the strange eye contact meant.* As if in answer to that thought it became quite apparent when his eyes penetrated right through to her soul and his weakness was exposed. Their patient relationship no longer existed and he could not justify keeping her as a private patient. They had clearly bonded. Bridgett knew how badly he wanted to get inside her head. Maybe this would be his reason for her to stay with him. She watched him deliberately trying to create a disease to find the answers he was seeking. Which she was not going to let happen. Martin was faltering more and more as she watched him jumping from subject to subject and pacing up and down. *For goodness' sakes,* Bridgett thought to herself, *what the hell is he doing?* She knew that the only assessment that could be validated was hidden behind her wall. *Perhaps that's what he was after and nothing more.* The only health issue confirmed and diagnosed was that of exhaustion; which is an acute medical problem and not a chronic disease. The physical assessment of Bridgett remained unchanged, as Cr. Creed had concurred. He went through every test, and all results were still found to be normal. He offered no concerns of any serious matters. His gift was certainly not working. Bridgett became frustrated with him by releasing his desires to maneuver her mind. So, whatever he was up to was taking a toll on their relationship. Bridgett's thoughts kept getting deeper and deeper. *Maybe, just maybe, he thought he could crack the nut.* Perhaps allowing his emotions get too close there became a realistic fear of confronting the unimaginable. Bridgett was well aware that Martin couldn't handle failure very well, and was remotely nowhere near removing one block of the wall. Was this a charade he played? Was there really any true bond between them? Maybe she was too much of a challenge to take on, which produced the smart ass grin from taking place upon her face. Then the real Bridgett stepped forward. "Martin you seem to be holding back something you want to express. I hope it has noting to do with me." Bridgett remarked, "After all, "I've been perfectly honest with everything involving our relationship. Is there something I missed that would appease your challenge?" Martin made an abrupt turn looking straight into her eyes and soul saying, "What challenge are you referring to Bridgett?" And she replied "The fact that you have never failed with finding that diagnosis which all other physicians missed." "Maybe I wasn't your best candidate for finding that mystery illness." Bridgett saw the anger in Martin's dark brown eyes unfold speaking directly into her face while kneeling at eye level stating. The only thing botched, Bridgett, was allowing me to trust you. This lack of trust, and the wall you hide behind has everything to do with your happiness, not your insomnia or exhaustion. We both know what it's about and until you decide to come clean and open up you'll always miss that one person who could change your life for the better." Bridgett didn't like where he was going. The accusations about purposely trying to divert the love and closeness obtained within relationship was all she did want, more that anything in life. How wrong, how far away from the truth he had perceived her true intentions. Bridgett wanted more than ever to belong with that special someone. However, thoughts of divulging the secrets guarded behind cement walls where the dark dreams lay hidden, only exasperated idiosyncrasies of other ideas. Those of which were categorized as being crazy. Bridgett was forced to relinquish all desires of building their relationship. The trust issue had not been acquired to its full capacity; which left Bridgett's desires of any further attachment to Martin come to a final decision. There was more time needed, and it was obvious he wasn't willing to wait any longer. Stopping the pain of heartbreak before it became full impact, Bridgett brought humor into play.

What about a magic pill, Martin? Just one magic pill. Maybe then I can pull my life together as you wish." Then she laughed, because crying was always suppressed. He heard every word that was said and avoided the conversation by staring out the window of his office. Apparently not finding the joke amusing. Remembering back how she felt with the government and their jokes, opening deeper scars. Then as a fly caught in a spider's web, he was lucky to get untangled and set free before becoming part of her food chain.

With her own laughter thrown into dead silence; showed Bridgett that there was no humor left with Dr. Martin Levy. His eyes fully focused into hers, he finally spoke, "What are you feeling now Bridgett? Did the laughter ease any pains? Do you want to share it's meaning with me? It must be immensely intriguing, because that's how I find you." The gavel came down upon the desk softly but indicative of making a thorough statement. He was now justifiably uneasy in her presence as well as intimidated. Bridgett stared back into Dr. Levy's eyes with a straightforward acknowledgment. "I'm quite flattered that you find me intriguing, however, I am a pessimist, and right now, I just don't convey the same attitude as you." After that statement Bridgett knew that the deep hole in her heart could never be filled. She reconciled to the ugly but surreal truth that there was no entrance for happiness only an exit as long as the dark dreams existed. Maybe they would slowly and painfully engulf her as a macrophage engulfs diseased and dead cells within the human body. Now as a physician and not just a new friend, the diagnosis introduced—sleep deprivation. Dr. Levy explained to her in no layman's terms that sleep deprivation could be related to brain disorders. He continued on to lecture that after periods of extended wakefulness or reduced sleep can cause neurons to malfunction, visibly affecting a person's behavior.

His final act of humanity came to a close that day as she was referred to another physician on the list. The credentials after his name was, "Psychiatrist." He was a f—ing shrink. Bridgett gritted her teeth making all attempts to stop the volcano from erupting. How dare him send me to a shrink to appease his inadequacies. This was a get back for sure, but it didn't stop the journey of discussing the wonderful background of Dr. Zong. He was the best psych in New York with name dropping added to spice it up, as if she really cared. Like drawing nectar from blossoming flowers by using his best efforts in selling his idea, Bridgett made no response, she excused herself to the ladies' room. Under the light above the mirrors, Bridgett took out the list of physicians from the Secret Society. Well, lo and behold, what do we find? Well, it's Dr. Zong. Laughing again loudly with a non caring attitude she spoke to the mirror, "How stupid I am to think that the next physician on the list would not include a shrink. Every physician' had assessed her except for the shrink. That was the trump card. Bridgett now understood what had transpired during their so called relationship knowing Martin couldn't pull anything else out of her he counted on a shrink to do his dirty work. "Damn him!" Bridgett continued to yell loudly. "*Am I really nuts? Does he really care? Were all our times spent together nothing more then friendship? Did I pose a threat to his credentials as a challenge that went out of control because I became a personal interest?* All these thoughts became rambled ideas until finally she took back herself control. Bridgett went back into Matin's office knowing it would be the last time. He was sitting on top of his desk as if in anticipation of her return. The communication continued where they had left off. "*I know that the referral to see Dr. Zong offended you, and I'm sorry, Bridgett, but he's going to be your last hope. Don't you understand, there is nothing physically wrong with you, except for sleep. I can't help you with your sleeping problems, he can.* Then he began pleading as if she was his wife and was leaving him for good. "*Please for the sake of our friendship, Bridgett, do it for us. You know you can't toss everything we've shared together away, and you do know how I feel about you.* Bridgett stared at him, thinking, *Do I even want to deal with this drama any longer? Being dissected like a frog in biology, as he waits for information that may be of value to him. What type of relationship would that entail.* Bridgett would have to think really hard about that question. Right now, she just wanted to put out the fire saying, "Yes, Martin, I will give it one last hurrah!" He then approached her close with arms stretched out for an embrace. This time it was different, and Bridgett backed away quietly closing the door behind her—like the chapter of a book.

As Bridgett hit the streets hailing a cab, she body slammed into someone else. In New York, if this action occurs someone needs to apologize or else it's not very pleasant. With that thought in mind Bridgett took it upon herself to apologize. The person turned out to be Maxine, a girlfriend of one of her colleagues, Bill Grahm. He was a very dedicated scientist, and Tom gave him a position at one of the labs. Bridgett greeted Maxine first with an apology, and then as a friend. She asked about Bill, and apologized for being so distant with their friendship, and explained the reasons; leaving a promise to get together soon and more often. Maxine looked at Bridgett and their eyes locked. Maxine stated, "I'm sorry. I know who you are, its Bridgett Montgomery, right? However I have no boyfriend by that name, and I have never met you personally at all. I have seen your face published everywhere. Your very popular, to say the least. Good for you, finally a woman that makes science headlines! However, I find that you have me confused with someone else, after all you said you hadn't seen "me" for quite a while. Maybe you just need a vacation to get away from all the publicity." According to Bridgett, if you can connect and lock eyes, then there was truth behind what was being said. However, the sound of the honking cab broke the trance and gave Bridgett reason to escape. Once in the cab, her eyes swelled up with tears, and she broke out crying. Arriving at her safe zone, she looked into the mirror, and there were no changes that were noted. Then she reevaluated the experience and wondered if Bill and Maxine had split, and maybe that was the reason she was upset and being so spiteful. Bridgett thought to herself, *When I see or hear from Tom, I'm going to tell him how Maxine treated me! Acting like she didn't know me, and leaving me to think that I'm off my rocker, and I need a rest.*

Bridgett was now infuriated at Maxine's attitude. All Bridgett wanted to do was pour a glass of wine and unwind from the incident of Maxine. Remembering the appointment with the shrink in the morning; she downed the remainder of the wine.

That same week, another occurrence took place that shook Bridgett's life even more than the Maxine incident. The episode happened at her favorite deli owned by a big heavy man by the name of Mr. Lamata. His broken dialect with conversations and jokes made shopping fun, with that certain sprinkle of delight. Entering the deli, she was greeted with his super wail of, "Hey, my lovely lady, what is going on here today? Are you playing a joke now on me?" This weird conversation came as a total surprise to Bridgett, because she didn't understand his greeting; it was deemed out of character for him. Mr. Lamata also seemed a little bizarre that day as his laughter seemed to escalate in an abnormal fashion echoing throughout the store. He looked puzzled, and kept lifting up his cap to wipe his head. This started a repetitive motion. Finally, he stopped long enough to remove some of his anxiety by saying, "Okay, Ms. Bridgett, you fooled me today. A good one. You like to get me back with some jokes? Okay, I understand. I deserve. I know, I deserve, but I not figure the joke, and I give up, Ms. Bridgett. Okay, I don't know who is who. You keep a good secret from me."

"This lady she say her name is not Ms. Bridgett, so you are my beautiful lady, right?" Bridgett felt the trembling starting by viewing her hands shaking very noticeably as though she had Parkinson's disease. The atmosphere in the deli changed dramatically as the temperature plummeted to an Arctic cold, and a thick mass of heaviness empowered them making it difficult to breathe. While everything weird was transpiring; she forced herself to maintain a sense of stability in answering Mr. Lamata's questions. "Mr. Lamata, I don't understand where you're headed with all this talk." Bridgett glanced around the deli and saw only one other person in the store. Again Bridgett questioned him, "I'm not understanding your joke today." She watched him glance back and force from her to the other person in the store who was a female, and a foreboding came to pass. This uneasy feeling emanated from the woman's presence next to her. There became an ambiance of something evil also combined with a déjà vu. This brought forth extreme anxieties for Bridgett as she compared them with the onslaught of the dark dreams. Mr. Lamata's odd behavior was now understood as the lady turned and they were face to face. Their eyes met and became locked together, and the silence was deafening. There was a brief moment that came and went; where time and space were disconnected. It was as though a mirrored reflection of herself was being displayed. She could feel the sweat running down her face and down the back of her neck. The only sound breaking the silence was the pounding of her heart, thump, thump, thump, waiting for it to jump out at any moment. This occurance only took place on the aftermath of the dark dreams. However, Bridgett continued to hear Mr. Lamata somewhere in the distance, saying, "Introduce me now, Ms. Bridgett, so I will know next time who is who." That day would always be etched into Bridgett's memory. This was definitely an experience that was not of the norm and could not be erased over a period of time. Bridgett was left clueless and confused, wondering if the dark dreams had now invaded the awakened part of her life. All thought processes became numb and stopped. She couldn't produce the meaning of this new encounter or how to survive the engagement. The strangest part was when it seemed that her mirrored image was not at all affected by the discovery. The eyes bothered Bridgett immensely as they presented them to be totally empty; compared with the eyes of a *shark*. There was nothing behind them; *expressionless* was the only word that could pulled from the stare. The eyes gave the presence of being fake, not real. Bridgett's most steadfast characteristics was the belief that the "eyes were the windows of the soul," as told so often by many, and now she realized just how true that statement became surreal. Their eyes were still intensely connected, but it was the other reflection that turned away first, not Bridgett. The joke by Mr. Lamata was now clearly understood. Bridgett didn't believe for a moment that this person was a mirrored image of herself, lost and trying to reconnect. On the contrary! Her theory was that the dark dreams kept inside were now releasing their presence while being fully awake forcing Bridgett to make a choice.

She needed to either ignore them and pass them off as skepticism or base them on the vivacious magnitudes of a wild "imagination." Being told that she had quite an imagination, she prayed that this was one of those times. Perhaps Dr. Levy was right when he explained that sleep deprivation could cause brain disorders, hallucinations, neurons forcing to reconnect, and a gambit of other medical problems. This made Bridgett stronger as those thoughts possessed her composure, walking with grace and dignity toward the exit. Thoughts to herself were, *Everything is all right. It's just my mind playing tricks on me. It's because of the lack of sleep, I'm hallucinating, I know I am.* The scientist now took over as she stopped before exiting the deli and yelled, "Bye, Mr. Lamata! I hope you and your lady friend work out all your puzzled questions." The mirrored reflection snapped around to stare at Bridgett, and she snapped back with the best picture ever taken to mark the occasion.

Bridgett rushed back to her safety zone and placed a phone call to a longtime friend and colleague, Lincoln Stoneburg. Lincoln was also within the circle of MENSA, and held a degree with a masters in microbiology. However, he chose another route of expertise. He had a natural gift for photography, and his pictures were exhibited in many magazines over the years within many countries. His gig was now taking pictures for paranormal research programs. Lincoln never expected to hear from Ms Bridgett Montgomery, whose project of using stem cells and embryonic implantations became International. Through Lincoln's amazment of Bridgett's call, also rang out a huge question, "Why?" The anxiousness detected in the message prompted him to call immediately. Upon reaching Bridgett apologies flowed explaining the reasons for her poor friendship abilities; but was immediately dismissed. Lincoln understood Bridgett's personality traits, and the persistence of pinpoint details. Perfection was a part of Bridgett's OCD (obsessive-compulsive disorder). She began explaining the huge undertaking and responsibility that was involved at the time of the project, then interjected, "Winning the Nobel was the easy part." *Oh my God,* was Lincoln's thought as he stopped at that juncture. Once more it came to him with clarity, *If that was easy, what the hell was taking place now?* Bridgett also explained that Tom had a role within the program and was placed within direct charge of all the labs. However, the marketing process needed the involvement of the public, and she volunteered her services. This way there would be assurance that more labs would be opened throughout the country; while Tom enforced the guidelines of the program.

Lincoln was also a childhood friend of Tom's and respected his scientific knowledge as being a genius, for lack of a higher rung of the ladder. Then, as if someone just belted him in the abdomen a churning started inside his gut that gave reason to believe trouble was on the horizon. Bridgett elaborated on the story of a woman she met at a deli that was a reflected image of herself except for the eyes. He could hear a plea for sanity rushing forth that gave Lincoln an unprepared platform to stand on. He was trying to understand Bridgett's explanation of their meeting, and the importance of a picture taken before exiting a deli. She asked for his expertise in validating this photo, and that it would be sent via computer by the end of the day. Lincoln's main concern was now the realization of how deeply important this issue had become as desperate pleas of "Help me, please, Lincoln," gave him goosebumps. He felt that she had given him scant information leaving out the importance of a "mirrored image." Lincoln was now very concerned for her and said he would do anything that was needed. Bridgett's final communication to him was, "Let me know if this picture is authentic, and I'm sorry we did not share a happy day together, but I'm hoping for a reunion soon. Also adding "What ever you find, good or bad, don't tell Tom!" Lincoln sealed that promise with a pledge from the oath they took at MENSA, then the phone went completely and silently dead. He stared at the phone as if a paranormal experience had just occurred. Lincoln was positive that more was hidden within that call then just an SOS from the circle of friendship. There was also an absolute need for rescue attached bearing great significance. Lincoln would critique the picture to the best of his ability as he would with all paranormal groups. Bridgett's theme about winning an award for best picture of the year brought back the humor that was part of her personality. However there were suspicious concerns regarding Bridgett's outrageous story. But, when the picture arrived, spontaneously the scientist kicked in to action as the picture visioned gave rise to something evil that made its appearance—the dark shadow figure. A different atmosphere started taking place, a noticeable change in temperature occurred as Lincoln began shaking from the cold. A terrible stench took place from everywhere making him gag and vomit profusely. The darkness that came closer and closer to him ascended as a huge cloud representing pure evil. Bridgett's picture was very authentic. Being a special friend and sharing a bond left a very scary thought involving a life and death struggle. Since the phones were dead he had no way of reaching Bridgett to offer any warnings. Lincoln continued to struggle with the putrid stench, followed with continuous vomiting. The heaviness upon his chest now developed into a shortness of breath. Lincoln knew that everything he was going through started when Bridgett called. Lincoln's only thought was the undeniable fact that she was in terrible danger. Then the darkness enveloped him leaving no signs of light, just a pitch black void.

It took everything for Bridgett to seek out Tom, but when he opened the door and their eyes met, she knew he still cared. Tears streamed down her face, and a loss of control with emotions took over. Tom, being startled by this new characteristic, realized that Bridgett was in a terrible dilemma. He took a step back to make sure it was her. Then he held her tightly, almost crushing her into his body, like an overprotective parent. Tom whispered quietly into Bridgett's ear, "Do you need a glass of wine, my love?" Bridgett answered quickly, "Tom, are you for real? Like, yes, I sure the hell do." They both moved over to sit on the couch, and there was complete silence, at which point she looked deep into his eyes searching for his soul. It didn't take Bridgett but a miniscule moment to know that he already knew what was transpiring deep inside. It was scary the way he could read her thoughts. She seldom said a word; he would just know. "Do you want to talk about it, Bridgett?" Tom asked. "Yes." Bridgett sobbed. "May I?" Tom replied, "I've always been in your corner telling you that I would never leave. So what are your fears, Bridgett? Let me help rid them for you." Bridgett began, "I' just want to start by saying how sorry I am for ignoring you. I was so wrapped up in my project that I lost everyone close to me." Tom tried to comfort her by saying, "Bridgett, I know. I understand. You don't have to explain to me. I know you better than you know yourself." Bridgett looked up at him and asked, "Really, Tom?" "You know me? What if I told you about my dreams, Tom? The reason I would wake you up with my screaming. Do you remember those days, Tom?" "How can I forget them? Tom replied. You were very afraid, and I was terribly worried. It makes me sick inside when I'm unable to help you, the one and only person that matters in my life." "Really?" she retorted. "How would you react if I told you a secret; something that happened to me at Lamata's Deli today. Something that points to a mystery, that I truly hope I'll never have to solve. She could see Tom's eyes sucking in all that was being told as she continued with the mystery. "Tom, I swear this person, this expressionless thing was the mirror image of me!" Tom interjected, "you mean like an identical twin image"? "Yes, exactly that, and I went as far as taking a full-length picture of her and sent it to Lincoln for verification of my sanity. How about that, Tom? Do you still believe me or are you ready to give me a label?" Tom got up and poured them both more wine. "You know, Bridgett," Tom said, "I'd believe you if you told me you were abducted by aliens." Bridgett laughed. "I wish it was that easy." Then she tried to remember the last time laughter meant joy. Yes, there were other times but they seemed to be always connected with Tom. Somehow any pain was removed by the introduction of laughter. Tom always found a pathway for her to escape. Now, she knew he was ready for battle by announcing, "Fire away!" "I'll need another glass of wine, maybe the whole bottle," Bridgett said as she handed him the glass. Tom laughed at her, "There's a liquor store just down the street. I can get all the wine necessary if it will help to free yourself from the tangled web that has you trapped." Tom watched Bridgett pouring the wine into a never-ending stream in disbelief, as he remarked. "Wow, I've never seen you consume that much wine before. You were always a light weight. I was the one picking you off the floor, after two, drinks. Now it appears you have grown a lot in my absence. Your still wide awake, alert and with no slurred speech after drinking an entire bottle of wine." "Well, maybe I'm not such a cheap date anymore." Bridgett remarked. Tom put his arms around her, pulling her near to him saying, "There's nothing cheap about you, Bridgett. You're worth your weight in gold and just don't know it."

Bridgett did not hesitate to talk about the dark dreams. Pointing out their origin starting around the time of the project. She continued explaining their habitual access into her life. When all was finished, Tom got up and said calmly, "Let's sleep on this tonight. I have some ideas that might bring some light into the picture. I can't stand watching you close to the brink of insanity. I know you Bridgett and about now your bets are placed on being crazy, but your not, and I'm the person that will prove it." "So are your sticking a label on me saying, crazy or maybe I am and you won't admit it." Bridgett stated. Tom looked deep into Bridgett's green eyes saying, "Nothing would ever make me believe that of you. I feel that your dark dreams or perhaps your subconscious is wakening something sinister. You've always been connected in some way to the supernatural, the paranormal realm. It was presented to you as a gift; while trying to conceal it from others. But, I believe in them very much, and I can sense there is a definite link between you and the dark dreams." "Why would you think that, Tom? What would be your reasoning?" "Bridgett, before I take you and kidnap you for the night and pretend you're all mine, I need to ask you something. It will be a very touchy, and delicate subject that I will need to question, but it might be the link that will divide you from the dark dreams. Bridgett agreed, while holding her breath. Tom kept hesitating and now there was reason to worry. This prompted Bridgett to force him to ask the question. "Hurry up and ask me, before I lose my nerve to answer, would you please? I want to spend time with you and not be interrogated," Bridgett scolded him. "Are you ready?" he asked. "More ready than you think," Bridgett answered. "Were you one of the donors for the embryonic program?" Bridgett laughed mockingly, "Of course, I did, Tom. What a stupid question. How would it look to other females if I urged them to donate and not me. This is my program, I believe in it with every part of my soul. How phony would that make me, if I didn't contribute to my own developed project.

This was the last conversation Bridgett remembered, as the wine finally coursed through her system. The phone ringing came as a startling jolt, compared to the dark dreams. She watched Tom pick it up and turned his back away from her while whispering took place. Bridgett became inquisitive due to the whispering and nothing more. Although it thorough amused Tom's thoughts as he believed them to represent a form of jealousy. Bridgett was worried that it involved one of the labs, and told him. Tom assured her that the phone call encased no dangers. She asked him what it did involve, and he chuckled saying, "That was Lincoln, about your picture taken at the deli."

Bridgett could hardly breathe. "Well, come on, what did he say? Hurry up. I'm dying to find out."

Tom continued with his chuckling saying, "Lincoln said it was the best picture ever taken of a menu." "There, now there's nothing more to worry about." Then he hugged Bridgett closer to him and never said another word. Bridgett lay furious as she distinctly told Lincoln not to tell Tom. Why would he turn against her, what were his reasons? Why didn't he just leave a message on her answering machine?" The biggest question that seized her every thought was, "Why would Lincoln sell me out?" Thoughts produced answers and the only one remotely associated with Lincoln's call was about Tom. She wondered if he was running his mouth again about reuniting their relationship, and Lincoln assumed it to be true. Bridgett pushed herself away from Tom's arms and slept in a fetal position the rest of the night.

The smell of fried bacon and fresh coffee awakened Bridgett the next morning. *Funny*, she thought as she was showering, *I don't remember having any dreams last night.* Bridgett couldn't wait to tell Tom the good news, after showering and dressing. Following the wonderful aromas coming from the kitchen, Tom's presence was absent. In his place were two of her favorite colored roses, yellow and red, and a note that read, "My dear Bridgett, I wanted to tell you last night that I loved you, but I didn't want the wine to interfere with any good intentions. I also don't want to put any more on your plate. I have some interesting work cut out for me today, which pertains just to you. Have a good breakfast, relax at my house, and I'll be back at six this evening. I love you, and now I'm headed on my mission." After reading the note she began to wonder what mission would be connected to her, except for the dark dreams. Thinking to herself, *How in the world could he help me with my dark dreams?* Bridgett did as Tom suggested—rested. However, her mind was at warp speed concerning marketing and opening more labs. These interruptions of thoughts left no rest or relaxation. She understood the necessity for embryonic donors. Without them, it would inhibit the life of stem cells to exist. Reaching out to donors was imperative, and sitting and watching cartoons was not productive. *Nice try*, Bridgett thought of Tom's idea. Her green eyes were wide open allowing all light to enter as it opened the doorway of reality to begin marketing the project. She would take time off later to spend with Tom. Anyway her abstinence continued; the rape factor had not yet dispersed. Yes, her thoughts agreed. *I'm forced to travel. There is no other way to get to the main population and answer questions. I need to get the attention of prominent people holding high positions. The project which was developed to extend the longevity of mankind; needed more donors and more labs.* Bridgett would have to shed the skin of a scientist and become a social butterfly.

Upon many travels, Bridgett found a spectacular place called the New Hampshire Lodge to hold conferences regarding the project. The lodge seemed to beckon her in a mystical way, as it exposed its location very prominently, even though it was hidden from view. It's massive size was removed from the main road by a few miles. *How weird was that? It was as though it was summoning the words come to me by the many trees swaying in the breeze pushing her toward the direction of the lodge.* Meeting one on one with the manager, Mike, she was taken on a tour of the beautiful, spacious acres displayed by a multitude of different trees with varies sizes and shapes, and their awesome colors displayed uniquely. There were clear fresh water brooks running everywhere throughout the property and easily found trails that would lead back to the lodge as a protection from getting lost. Mike gave Bridgett quite a bit of information about the lodge after the tour; which Bridgett found very interesting. He stated that the lodge was not open to the public, and strict implementations were placed upon each guest. This would include an invitation from the lodge with an RSVP attached. They excluded the lodge from magazines and travel agencies. There were very few staff changes; therefore allowing continuity for their guests. Bridgett's thoughts at that time stopped at the invitation part, because she had not been invited. Feeling embarrassed by the abrupt arrival, an apology was issued. "I'm sorry for the intrusion, I could see the lodge in the distance and I just fell in love with it. I wanted to ask permission to hold my conferences here." Bridgett continued, "I don't normally just barge in somewhere. I prepare my own invitations as well. So, I truly understand this inconvenience, and I'm very sorry. I had no idea. I just saw it." Bridgett's stupidity continued to pour out of her mouth, lacking any control. Finally, Mike intervened, "Ms Montgomery, please allow me to address this issue." Bridgett, still feeling like a babbling idiot was more than glad he decided to take over. "We were expecting you. We received an invitation asking permission to allow your conferences to be held here at the lodge. We already received the agenda with reference to each month, day, and time. You seem to have a full calendar of events outlined." Then he took her to the destination of the dining room where the conferences would be held. Bridgett found herself again in another moment of time suspension, as every word Mike said was magnified and was recording at a slow pace. The lodge had beckoned and summoned her presence this was a fact. There was no reason that could be acknowledged for what just transpired. The lodge could not be seen for three miles from the road. This alone shed no feasible answer but "paranormal"; it was not a coincidence. There were no words and no thoughts that could ft the description of this new experience that just transpired. *Could it have been involved with the Secret Society?* The only thing Bridgett understood were the feelings that came flooding forward upon entering the lodge that day. She could feel it breathing as it came alive by her presence. It brought its own entities shielding and protecting those who were invited, and entwined itself with Bridgett in perfect harmony.

The staff was very personable, almost bubbling with excitement as observed during the conference that evening. She was being appreciated for bringing not just monetary rewards *but new life* into the lodge. Bridgett couldn't help but overhear the staff's enjoyment while working expressed by their happy greetings combined with laughter and smiles. They brought no negativity just positive joys beaming from all directions within the lodge. Their attitudes remained the same when Mike invited her to become a private guest with her own key to room #24. The members of the lodge became very protective as they screened all her calls and only took messages. They never allowed anyone to know she was there. Bridgett always had room service unless otherwise noted, and there were security guards located at every entrance and exist who carried weapons, exhibiting no facial expressions, and never talked to any of the quests only pointing toward the desk with questions. Bridgett had her own body guard outside room #24 and allowed no pictures to be taken within the lodge. Mike knew of the spotlight that surrounded her; combined with the many important and powerful figures that the lodge sheltered. Bridgett felt more like a celebrity than just a scientist when staying there, as it became her sanctuary.

Lacy Pittman, the name fit for a simple person, was not complicated, but her personality did not coincide with those outward characteristics. Apparently she had no desire to impress anyone, as noted by her drab attire and mousy dishwater blond hair. However, Tom saw something else in Lacy, a bright light with a well pronounced aura. The colors of yellow and blue indicated the right person was chosen for the job position opened. This in and of itself made up for the non impressionable persona. It also verified his innermost feelings that Lacy would become a valuable person in Bridgett's life. Tom had heard through friends at MENSA that Bridgett's health was questionable—a state of physical exhaustion. He knew nothing about Bridgett's poor health and issues concerning her well-being. Tom knew about the unrelenting back-to-back conferences that were necessary in opening as many new labs as possible. He was knowledgeable of Bridgett's pursuits with everything that contained tedious efforts and pinpoint perfection. Opening labs for the stem cell and embryonic implantations would make the OCD's even worse. Tom did not like being in the dark when it came to anything involving Bridgett. Now, his challenge would be how to help her prevent physical and mental fatigue without initiating any forceful confrontations. This was when he came up with the idea of hiring a personal assistant and confidant to unload all that was monumental with involving Bridgett. Taking this action upon himself was also risking any chances of them being together. Should he risk it? He proposed that if he didn't Bridgett would take a turn for the worse, and he couldn't endure that sick feeling in the pit of his stomach. His only ray of light would be that she would be too tired to think about this proposition. With thoughts of Bridgett's destitution in mind, Lacy was hired. Tom's initial inner feelings about Lacy brought many positive connections, and the urgency for her to start work was essential, and Lacy had no qualms about starting the new position as Bridgett Montgomery's personal assistant.

He felt confident with Lacy's abilities, intellect, and even her straight forward approaches. Tom knew that Lacy would be the female of choice for Bridgett. Wasting no time at all, he sent her to the conference that was now underway at the Concord Lodge. Tom understood that there was something about the lodge that captivated Bridgett, calling it home and her safe haven. She had invited him on many occasions, but Tom never took her up on any of the invitations. He thought it best not to tread upon her space. In the back of his mind, the lodge somehow did not mesh with anything that was considered to be good. Therefore, he would need outside help giving Lacy a push out the door and on the way to the airport; with the Concord Lodge as a designation. There was still enough time to get there before the conference ended. Tom's guilt of Bridgett's crisis made him believe he could still be that white knight in shinning armor that she was positive existed. Because of the highest priority bestowed upon him by his boss, Bridgett, needed him to stay focused. No mistakes were tolerated. It was not an easy job, but there had to be a reward in the end. Tom held back a lot of information that wasn't relevant to discuss with Bridgett. He basically chose to handle a lot of issues by himself to prevent any unwanted worries. Tom knew that there was enough on her plate already. His recommendation to Lacy upon leaving that evening were the ways to deal with Bridgett's strong characteristics and personality traits. This gave Lacy an initial start in knowing about her boss. The new position as Bridgett's assistant required important tips that down the road would come in handy for Lacy, and would someday be very appreciated. Upon boarding the plane he kept throwing in more information for Lacy's benefit. His most important advise the first evening they would meet would be to sit quietly, perhaps way in the back unnoticed; while listening intently as Bridgett explained the process of embryonic implantation with stem cell implants. Even though Lacy expounded upon the knowledge of the project, Tom explained that Bridgett was unpredictable. Sometimes there would be new research introduced into the program. Lacy turned to Tom and, very childlike, asked him, "Tom, can I tell you something about myself that I think is very important?" "Of course, you can," he replied. "I want you to realize that Bridgett is very special to me and to many others. Her theory has been followed every inch of the way. I completely understand the program and hope someday that she will feel confident enough to allow me to give the conferences so she can take a break. Until that time, I will do what is expected of me, and if something goes a ary, I will call you." "I could never match my knowledge to that of Bridgett's and the program. However, I feel confident enough to respond appropriately to answer questions presented after the conferences. This would give Bridgett time to actually meet with clients to market the program. One more thing I just thought I'd throw in, I too am gifted. I feel others' pain and emotions, and I feel that Bridgett is in need of my assistance, and I have to get there now." Tom watched as Lacy's plane became nothing more than a metallic object high in the sky.

Tom sat and pondered as to what had just transpired between him and Lacy. But, all he felt was peace and a sense of security for Bridgett that she would find in Lacy. Tom saw her as another of the many hidden people that were beginning to surface since the gateway of scientific and medical advancement was enhanced by Bridgett Montgomery. Tom was very moved by the fact that he could do something genuine for

Bridgett without her direct permission. The extra blessing thrown in with Lacy (being gifted) was justification that all angles of Bridgett's dilemma were covered. Bridgett was thought of as his entire world, and would protect her at any cost. The only menacing factor that left any major concern was the dark dreams. That intricate part of her life would be dealt with on an entirely different level. There were no set rules forcing Lacy to cope with them; it wasn't part of her work position, just Bridgett's assistant. Tom would die before ever allowing the outside world to know the secrets that were locked away, and the true wrath of Bridgett's personality. If Lacy wanted to proceed close within Bridgett's secret world, then she would learn everything through her mentor. Tom never withheld any information from Bridgett, and now picking up the phone, there would have to be a logical explanation for the action of hiring an assistant without consent. All he could hope was to avoid fireworks. Though there wasn't much to say about Lacy, he knew that Bridgett would understand the reason she was chosen. Knowing that Lacy was different and gifted would be approval enough, he thought. Bridgett's forte was always seeking out those individuals who did not fit into the society's realm of being "normal" that were cast out for being "weird" or "strange." He hit the jackpot that night as Tom could hear Bridgett's excitement over the phone and concurred with his idea of hiring an assistant. Her voice also expressed anxiousness and anticipation with meeting Lacy. He sighed with relief that she was pleased, while wiping the nervous sweat from his face. Now, he could focus much better upon his own job, as his only worry, Bridgett, was in good hands. Tom would always provide the best for Bridgett, at any cost.

Lacy did as instructed by Tom. She stood in the back of the room and listened intently to Bridgett and fit into the norm of society residing within the conference. Lacy could feel the hairs on her arms standing straight up as goose bumps were prominent; while listening intently to her lecture. Lacy was sucked into Bridgett's web very easily, as the others, who never faltered from their focus upon her presence. She was now a part of the audience mesmerized completely by all that she entailed. Starting from her radiant and natural beauty to the intelligence of a woman who beat all odds empowered by a world of men with a monstrous profession, and still won the Nobel. Bridgett was an exceptional gift to society, and Lacy wanted to be included and accepted in this new world. Her beliefs and morals were shared with the involvement of stem cell implantations and embryonic implants and wanted nothing more than to participate in circulating this knowledge. Their goals matched head-on with good intent to keep the program on an upward spiral of success. Lacy now understood Tom's fears and hesitations with Bridgett as she grasped the qualities involving Bridgett Montgomery. It entailed a whirlwind personality and the attraction of a magnet to a meteorite. This night, Lacy knew, would be the first memory that would take possession of her heart.

Bridgett finished the conference and proceeded to mingle with all her guests. Something strange took place that evening, as Bridgett's eyes viewed a shooting star from one of the dining room windows. It had been so many years since one had occurred. Bridgett closed her eyes for a few minutes to make a wish, as understood by many believers. Standing quietly, eyes closed, she suddenly felt the warmth of soft arms wrapped around her tightly. Slightly startled, and opening her eyes hastily, she was deeply gazing into the eyes of, who she believed to be, Lacy Pittman. Bridgett knew somehow that this was the assistant Tom had hired. She could sense something quite different that distinguished her from most females. Lacy had the essence of empathy, and tenderness linked to that of a nurturing mother. Bridgett loved the fact that Lacy didn't hide her feelings and had a very strong persona. She was outgoing as a flower child handing them all out to everyone, saying "Peace," and yet there was another side to Lacy's personality that said, "I'll get what I want, no matter what it takes," and added to it was an "I don't care what you think of me," attitude. Bridgett loved it. Lacy was spectacular in her own right as she displayed a large part of her own personality. There was no doubt in Bridgett's mind that Tom, once again, came to the rescue.

The conferences were always lengthy, and many times Lacy would have to finish them because of Bridgett's fatigue. Lacy, now understood how grueling the conferences had become, as more and more people sent letters for invitations, to hear her speak; Lacy was happy that the program was getting through to so many people. She felt that Bridgett's exhaustion was because she was not taking care of herself properly. Lacy was aware of all the years Bridgett had worked on the project, as did Tom, and now it was time for her to reap the benefits. Instead it was taking a toll on Bridgett's well-being. Lacy involved herself with all marketing duties and realized how little time Bridgett had to socialize with her guests. It came to the point where she blatantly told Bridgett, straight out and sassy, "I am going to start speaking for you, and I want you to just mingle with all the politicians and grace them with your presence like you do so well." Bridgett grabbed Lacy and hugged her. "Lacy, you are a special gift sent to me. I trust you handling my speeches with accurate literacy of the project. You learn fast, and I have heard you speak many times in my place when you weren't aware. I trust you!" Communication between the two were resolved, and Bridgett was getting ready to greet her the guests.

Lacy could hear Bridgett screaming and yelling while asleep far too many times. Springing up fast to try and wake Bridgett was the norm. This task was accomplished by yelling her name loudly and repeatedly. However, this time, it was working. Lacy was very scared! The only thought that came quickly to mind was to grab Bridgett by the shoulders and shake her until she responded. In the process of shaking Bridgett trying to awaken her, Lacy noted that there was no weight to her; it was like shaking a rag doll. Once she became semiconscious, with responding to her name, Lacy started the questioning. "Are you having a bad dream, Bridgett?" Bridgett recognized Lacy's voice and replied very faintly, "These are not just bad dreams. They are called dark dreams." Bridgett stated quite clearly. "There must be some explanation for them, maybe something medical?" Lacy asked. Again, Bridgett responded, while lacking energy and barely able to speak, "I have been to medical physicians, Lacy, no one has an answer." She had watched countless times the bizarre episodes always rescuing her from something unknown. These dark dreams, as Bridgett called them, always left her boss in a terrible state of mind. There were times when Lacy found Bridgett awake for days, using caffeine pills and countless pots of coffee to stay awake. She always asked her "Don't you ever sleep?, Why won't you allow yourself to sleep? Then the nurturing mother took over. You can't continue with all this work without sleeping. "Why are you so afraid of these dark dreams?" Bridgett gave a short synopsis of the dark dreams. "They won't allow any normal dreams to enter, just dark, scary, ugly ones that have entities that attack me whenever I close my eyes." Lacy was shocked to hear this, and further asked, "Is that what they do, Bridgett? This is the reason you never sleep?" Lacy knew the answers already; she just needed Bridgett to confirm them. However, this didn't stop Lacy from digging even further. "What are the dreams about that you claim are so horrifying?" Lacy wasn't expecting the answer she received from Bridgett. "I don't understand anything about them, I've been trying to find their origin and existence, but they remain inconclusive." "Bridgett," Lacy said her name sternly, "I see what they do to you when they leave as you struggle to awaken. Why are you experiencing all this chaos, of these dark dreams? Can't they be stopped? Why do they exist? Where did they come from?" Bridgett just stared up at Lacy and looked deeply into her huge blue eyes as she spoke. "I can't offer you any answers . . . I don't have them." The dark dreams never allowed anyone to wake her, but Lacy wasn't aware of that fact. She viewed Bridgett's face that night while brushing her long black hair and saw the beautiful green eyes that once held the twinkle and sparkle of a starry night, appearing as stars burning out somewhere in a nearby galaxy. The prominent dark circles with puffy eyes appeared as though Bridgett had aged greatly, surpassing thirty-three years. Her renowned enthusiasm and energy were depleting rapidly, and knew she was running on empty. Lacy's concern went even further than employee to boss. She loved Bridgett very deeply but never embarked upon that discussion. All Lacy could do was continue being supportive, holding tight to the nightly vigil while Bridgett's journey continued with the dark dreams.

Lacy became all too familiar with the dark dreams and was also affected by them. The gift for feeling the pain of others, whether physical or emotional allowed her to experience many things the dark dreams brought. She could not feel the full blunt of them, but only touching the surface made Lacy unable to endure them. These intruders in the dark dreams, she believed, were trying to break Bridgett's spirit. Knowing the strength and character Bridgett carried left her unable to fathom the hell that was taking place. Lacy held Bridgett close, all that emanated from her body was an icy coldness. She felt her skin wet and clammy turning the color pallor, etched with Grey, indicative of someone near death. Lacy wondered how much longer Bridgett would live, as the dark dreams continued to the suck every bit of life from her body; leaving her weak and frail. The concern of endangerment that they imposed upon her awakening led Lacy to surmise that soon Bridgett would be consumed by whatever the dark dreams symbolized. Until then, under her care, she would continue as the protector; which held no boundaries. Lacy's empathy and vigilance was all she could offer. She found that Bridgett seemed comforted when she would reiterate over and over, "Everything will be fine. Your stress is causing the dreams; which will leave when your struggle for the program becomes a reality to the entire country and all labs are opened." Bridgett smiled and played along with Lacy's philosophy, even though Bridgett was well aware of the ploy that she believed was a calming device.

Lacy remembered quite clearly the day she and Bridgett were to attend a conference at the Hampton Concord Lodge. They were about an hour into their flight, and Lacy took vigil as usual while finding Bridgett asleep. This was abnormal for Bridgett; she knew her obsession with staying awake, refusing to sleep and watching the struggling take place. This frightened Lacy, and quickly tried to keep her awake. Lacy began shaking her repeatedly until she responded. Bridgett opened her eyes and looked at Lacy, saying, "Why aren't you letting me sleep? You were the one telling me I had to, now you won't let me. What's up with all that nonsense, Lacy? I'm allowing the dark dreams to take me. I want to see why they are leading me back to David. They keep showing him, to me, Lacy, and I need to know why." Lacy asked excitedly, "Do you really think you'll find out the reason, Bridgett?" "Yes, I do," Bridgett replied very adamantly. Then to Lacy's surprise, Bridgett stated, "I believe I'm getting close to their significance in my life." Lacy knew that David was a large part of Bridgett's life at one time and was beyond delight to hear the adamant statement. Just hearing the strength of those words gave Lacy hope for once, instead of fear. Lacy also felt she had a right to know the truth of the dark dreams, as they made her part of their existence also. These simple words of affirmation to many would not generate any exuberance, but to Lacy, it was like being a little girl at Christmas, and receiving something special she had asked from Santa. With a trust of friendship that was bonded between the two of them, Lacy allowed Bridgett to sleep.

Lacy spent an entire year on tour by herself, while Bridgett and David became involved with each other. Bridgett emphasized to her that there would soon be a commitment made by David, and then she slowly started opening up about their entire relationship more and more. Lacy's wish for Bridgett was to be happy. Although never meeting David, the experiences shared by Bridgett proved that he did indeed exist. Bridgett's great love for him was felt by Lacy in great detail, as their intimacy was shared as friends. She had announced on one particular occasion that David said he had something special to announce on their first-year anniversary, leaving great hopes of a proposal. Lacy was ecstatic for Bridgett and was intrigued to meet the man who swept Bridgett off her feet. It was hard to picture anyone good enough yet she chose this man to be a partner in life. Knowing that Bridgett represented truth and honesty, Lacy hoped that David believed the same. Lacy had always surmised that it would have been Tom in the end. But, Bridgett never mentioned his name, and she didn't either. Tom had his own dislikes about Bridgett, and the only one that stood out the most was "being gullible." He told Lacy that Bridgett's philosophy was that everyone was truthful because they had no reason to lie, and there was goodness found in every person. Lacy was hoping that Tom's opinion of Bridgett was not correct. Maybe there was a definite link between Bridgett and David that was unbreakable.

Although Lacy's true validity of Bridgett's happiness would not be a proposal; but finding the means to an end with the dark dreams. As long as they existed embedded deeply within and tucked away, there would be no happiness with whomever Bridgett chose. As the flight continued to its destination, Lacy remained vigilant.

Bridgett wanted the dark dreams to lead her to David; therefore, she would have to fall into their ghoulish grasps. Sleep arrived quickly as she was easy prey due to sleep deprivation. Soon she was swept up by a gust of wind, headed down an endless dark tunnel. No lights, no sounds, just pitch-black darkness. This was when the entities, the dark shadow figures in the dark dreams would reach out from everywhere and attack. As the darkness surrounded and pursued Bridgett, the idea was that they would lead her to David. *Had they tricked her? David was nowhere around.* Bridgett got angrier and angrier at the thought of their deception. *Take me there!* she shrieked. *I want in!* In the distance, Bridgett thought she saw a sliver of light. *God, I hope they take me there*, Bridgett thought as she focused on the light. Remembering all the ghost stories in her past, all Earth-bound souls were to "go to the light." However, many times she floated through the darkness and never saw any such thing. Bridgett never knew where they would take her; it was always on a different adventure. There was no floating yet, so the only way to the sliver of light was to start walking, hoping that the entities within the dark dreams would leave her alone and escape their presence. But luck was never on Bridgett's side when it came to the dark dreams, and they heard her thoughts. Soon she began floating or levitating upward without any self control. She hoped that they were thoroughly amused and put her down, so she could run to the small silver of light exposed. As her body floated, it started taken on great speed, like an F5 tornado as it caused a twirling motion with various twists and turns, remaining at top speed. Bridgett didn't take fast rides too well, and with all the different movements of flipping and flailing around, she began to vomit. Everything was worth going through, Bridgett thought, as long as they took her to David. Bridgett knew this dark dream was all about him, and she needed to find out why. *Did the dark dreams have any connection to him?* The price of admission to have your fortune read was that she was now their personal property.

The attack was underway as all the black shadows figures grabbed her from every direction. Bridgett felt her limbs being pulled from many and all different directions. She no longer heard the usual, thump, thump, thump of her heartbeat. This led her to believe it had finally stopped. She surmised that death had finally came as desired. The small amount of light that came through the darkness showed the outlines of the shadow people, and it was not beneficial. They wanted her to view the tools of torment that soon would be used upon her. There were long instruments resembling needles and each came with an attachment of glass like syringes that dated back to the dark ages. She viewed the multitude of shadow people as they stood above poking them into all areas of her body. In the small amount of light she witnessed the huge gaping holes the instrument of torture left behind, and fluids seeping from them. They held tight grips, allowing no way to escape, while her mouth was being stretched open as wide as it possible. Then she felt liquids being poured down her throat. The fluids contained lumpy substances that had no taste. Still stretching her mouth open the liquids continued to be poured with more amounts of thick lumpy substances that emanated putrid and foul odors, and they would close her mouth after each amount. This prevented Bridgett from vomiting out any of their liquids and forced her to swallow her own vomit. They kept pouring more and more while picking up the speed of pouring, not allowing her one breath in between swallows. Bridgett gagged but couldn't expel anything that was swallowed, and the stench of the substance was unbearable and induced more vomiting to occur. She continued to swallow her own vomit, along with the fluids. Bridgett reached the point of suffocation as she never got one breath between swallowing, and slowly ceased to breath. Still somewhat alert, and with the tiny amount of light still available she watched in horror as the huge old fashioned needles continued to poke huge holes into her body, as observed by a massive amount of blood covering the entire surface of her body; resembling a murder scene. The huge amounts of liquid turned into sludge, it continued to be forced down her throat, but making it impossible to swallow. Her throat was closing, there was no room left to pour any more fluids. Bridgett felt the earthly existence leaving her body; the struggle to survive was gone. It was the only decision to make that would stop all the torment as she gave herself to the dark dreams and their origin. How foolish to struggle in torment just to find David, she thought. Then she felt her body started to float again within the blackness as the change in the atmosphere was making her more alert. Bridgett did not feel any further torment within the confines of the silent black abyss, all she could do was wait for the next step to occur. Being left without any control there was nothing more to do but wait. Suddenly a familiar voice was heard, a human voice that she recognized as David. Then his face appeared, and Bridgett wondered if he too was being held captive. *Maybe he was there to be her white knight to destroy the entities within hell. The dark shadows that took satisfaction indulging at the joys encountered while the tortures were inflicted.* They read her mind once more and mocked her, as their laughter bounced everywhere within the darkness. The dark shadows spoke in many different voices and different tones continuing to use her as a play toy until they became bored. Then suddenly everything stopped, and there was complete silence—like silence cancer grows as she heard David's voice somewhere calling out her name repeatedly from the forgotten area of the dark dreams. They did keep their promise!

The lodge took on a life of its own that evening, one that would never be encountered again. It was recalled to be exactly one year to date. Bridgett felt awful that the lodge did not receive the credit it deserved. Apparently, they knew how to play the most dangerous game imagined, and she would become just a discarded player. The attention that was lavished upon her was quite impressive. A thought emerged as Bridgett recalled the Secret Society's carte blanche offer to assist with any and all conferences. They promised to maintain a flow of powerfully important individuals to attend that would benefit the flow of opening labs. With great attempts to accurately recall the conference that special evening, it was noted that there were new and different people attending. During the well prepared information relating to stem cells and embryonic implantations all eyes focused upon Bridgett as being the main attraction. However, as she found out later that evening, her best admirer was David Bradford, Senator of New Hampshire. Bridgett felt his thoughts as they burned right through to her soul; while his dark brown eyes slowly and seductively undressed her. She picked him out through the multitude of people present that evening. Their eyes made direct contact from across the room, and Bridgett knew there lurked a reason. The tap on the shoulder came first, as Bridgett continued playing her feminine role, as in the lab slowly and methodically turned toward the direction of the tap. Their eyes continued to seep and mix, which pooled together as time stood still, allowing them to absorb one another. The vortex was closing around them, pushing them together perfectly to fit into a puzzle. Bridgett seemed to be melting into a pot of déjà vu. The dark dreams continued to bring forth David, as she noted absences in their relationship deeming it to be very vague. Bridgett didn't understand why the encounters with David were barely remembered. *Maybe*, Bridgett thought, *my subconscious is trying to force me to forget about him.* The dark dreams were leading her toward something important. Many adventures took place between them as a couple united in love, but they were making her solve the puzzle. Bridgett understood that if she didn't resolve this task, more and more of herself would be lost to the dark dreams, and the result would be the inevitable . . . complete insanity.

Bridgett's thoughts returned suddenly, remembering everything needed to complete the task thrown at her by the dark dreams. It was as though her subconscious finally woke up and kicked into gear to retrieve anything and everything they had to offer. David Bradford was in his early to middle forties, a gorgeous man and statuesque in the way he carried himself; his style of clothing was impeccable right down to the seductive cologne worn. The sex appeal that attracted Bridgett was so great it was unimaginable. His black wavy hair held patches of silver that embellished the sides and a large part of the front that made him look distinguished for his age. Bridgette nicknamed him "the hunky skunk." He had huge brown eyes, as dark as the dark dreams, but his were enchanting and seductive while offering an introduction of suspense.

They set off a mode of twinkling whenever he laughed, and David brought forth a lot of humor; which was a refreshing joy. They both had a lot in common and shared many of the same avenues of their lives. Bridgett knew they both were very vain individuals, as their physical attributes seemed to be a priorty. They were self-indulged with their relationship and had no world outside of them. Conversations were a learning period for Bridgett, as his knowledge encompassed 360 degrees of wonderful thoughts and ideas that were always intellectually stimulating. They enjoyed many of the same activities that stimulated their gain of vanity. There were many different adventures that David would find amusing, and Bridgett enjoyed them as well. The spontaneity, however, Bridgett had to admit came from David who always came up with a surprise. He said, "Whatever you choose to do or whatever happens on a daily basis, good or bad, is considered an adventure, and you have to flow with it. "It is, what it is," was David's motto. His feelings summarized this by saying, "Otherwise, your life will be miserable." *Point well taken and under consideration* was Bridgett's thinking. Everything they did together became a variety of events. Bridgett's feelings on that subject was in reality, as she always hated surprises, and knew that nothing would change that decision. Bridgett would never express that fact to him; because she felt it would place a damper on their relationship. Also, the truth was that there were great fears hidden involving a relationship. This was the first real love Bridgett had ever encountered. The truth be known was that her failure would be losing David. This gave Bridgett a feeling of fakeness made her feel like a hypocrite. Working in many women's groups, she would give talks about relationships. One of the main objectives was telling females, "Always speak what's on your mind, and don't do anything you dislike just to hold on to a man. Never show a man your weakness because they will take advantage of it." Yet to her dismay, she became one of those women. Bridgett's main goal still held within the dark dreams was knowing everything about their relationship, and she wanted to erase him from ever entering there again.

Once again waking up from the dark dreams, she came face-to-face with Lacy. "Well?" Lacy asked impatiently, "Was it good or bad this time? Did you find any answers to the dark dreams? Come on, you're killing me. Tell me, tell me!" Bridgett, understanding Lacy's excitement could only state, "I'll tell you later when we get home. I'm too tired right now." "OK, OK." Lacy begged. Upon their arrival back to Bridgett's apartment she spilled her guts about the dark dreams and the involvement of David. "Listen, Lacy," Bridgett stated, "just because no drastic changes are taking place right now, and it appears that the dark dreams aren't attacking me as violently as they used to, it does not mean that there's not something waiting. I can still feel them around me. There was a reason the dark dreams took me to David." "There has to be a reason they allowed me to take a piece out of the dream," Bridgett stated. Lacy expressed her annoyance of the dark dreams as she continued, "Why can't you just believe that they have finally released you, and it's all over? Maybe they just wanted to remind you of someone you once loved." Bridgett knew this statement from Lacy to be true. However, she understood the way the dark dreams worked, and they would never allow her any happiness. Lacy's voice could be heard across the apartment as she yelled at Bridgett in retaliation to her every comment, "No, Bridgett, I'm a realist, just like you were at one time." Bridgett cut the conversation short. She was exhausted from the torture of being in Hell just to find out if the dark dreams knew something about David. She couldn't believe that Lacy, of all people, could forget how savage the dark dreams attacked. Whether she called upon them or they kidnapped her made no difference. Torture would always be involved. Lacy had been with Bridgett when they occurred and even cleaned up the aftermath of their disgusting thrills. Bridgett wondered why Lacy was being so indulgent and understanding toward the dark dreams.

Sitting on the couch, watching TV and relaxing; the date and time was noted and drew her attention to it immediately. *Well, for heaven's sake*. It was indeed their first-year anniversary, as Bridgett felt a true relationship with David. Wow, if she hadn't seen the date pop up she may have forgotten. It had been a year, as her feelings about David matched no definitive time in the present world. Nothing mattered anymore to Bridgett; because her belief was that were meant to be together, as soul mates. Thoughts focused about him, and the phone startled her and broke the concentration. Call it a gut feeling or maybe there was more to this bonding they shared it was David calling as she quickly picked up the phone. Bridgett's hands still trembled, and her heart still beat fast whenever she heard his voice. David had a way of making her nervous and vulnerable; which Bridgett knew were danger signs for any female. These markers were not to be displayed; because then it would be known that he had the upper edge. She didn't want any man detecting her innermost feelings: those women always lost. The game was to keep men guessing about your feelings for them. Bridgett was about suspense and secrecy, and this helped in achieving many desirable situations. She was not going to allow David to open those secret doors or solving any mysteries. Never, Bridgett thought, would she show any weakness that would cause vulnerability. After all, she was the renowned strong and famous female scientist, BRIGETT MONTGOMERY, and he was lucky to have encountered her a year ago.

"Hey, babe, got some spare time for me?" he asked with a hint of laughter, which meant he was up to something. Bridgett replied in the same playful manner. "I have too many men banging down my door who just can't live without me. So, I really don't know if I have any time available." David was always quick with comebacks. "Well, I know one man who doesn't want any other men around to bang on your door." Bridgett was awed by that remark. David survived on spontaneity like no one she ever come across. He was the most unusual man to have the pleasure of meeting. The comebacks were great fun and always surprising, never dull. David completely entertained her intellectually. *It is a joy to have a man with brains and good looks*, she thought. Bridgett admitted that she never met anyone who could outwit her in head games. "Oh, so who is this mystery man who wants me all to himself?" she played. David seemed anxious about something; because he never gave in to the games easily. "Okay, you got me. Let's meet at the lodge. I have a special place to take you tonight. Real special," he continued. Bridgett knew at this point that David had remembered their one-year anniversary. "I'm on my way," Bridgett replied, trying not to sound too excited, while racing thoughts were entering at warp speed, and trying to catch her breath. She packed with speed and performance that prompted exact items to grab for the special adventure. Those that would accentuate her best features while keeping his attention. Bridgett's world revolved around David, as she tried convincing herself that it wasn't true. The depth of her feelings spawned an advantage for the other side. This new occurrence never happened before with any male. The worst part of the relationship was when she could feel herself lowering her guard and letting him through the brick wall. It was falling down brick by brick every time they were together. Bridgett wanted this new year to be fresh and new without any regrets of who gave in first, and somehow allowing the brick wall to fall. The one secret she would never tell would be about the dark dreams, and was willing to take them with her to the grave. She didn't even want to guess what would happen if they were exposed, as men were very poor recipients when it came to excess baggage. All David saw about Bridgett was the free spirit, and he enjoyed that part of her. He would never accept anyone out of control: appearing being weak and frail. That's why the dark dreams had to remain locked away. She had no regrets explaining them to Tom. *Why would I tell him and not my soul mate?* She began to wonder. The answer to that question was that perhaps it wasn't the right time. The dark dreams and David wouldn't mesh, she was sure. Bridgett could feel her independence slipping away, and found herself pulling back the reigns to slow down. There remained a fear inside of losing herself to David. She had fought her entire life for other women forcing them to stand their ground with men. Bridgett despised seeing women psychologically messed up because they gave themselves to the wrong man. Maybe she had no right to give those lectures to women; because she never felt what it was like, so how could she understand those feelings. Bridgett never really understood this diabolic plan and how easy it was to get trapped. If she gave in completely with signs of weakness, giving David the upper hand, she was very afraid he would leave and go on to another adventure. Bridgett hated the need to be with him constantly: never wanting him to go home or out of town. This was indeed the scary love she wanted to avoid. When a heart breaks, it doesn't always break even. Therefore, someone will walk away and lose nothing: while the other is met with insanity. Bridgett finally concluded that relationships always involved games. The significant part is that the better person walks away when the game finally ends. She would have plenty of time to deliberate these thoughts during the flight to meet the man that started a snowball effect by introducing her to love.

Bridgett exited the plane, and David stood out like a movie star wearing a sharkskin suit and a silk shirt with his hair adding the final touch. Bridgett exhaled deeply, as his presence took her breath away. She didn't know how else to explain this exuberance. Then, as if he didn't catch everyone's attention by his demeanor, there he was holding up a huge cardboard sign that read, "I'm the hunky skunk, and I love you." Bridgett could feel herself blushing from all the public stares. David presented her favorite flowers, yellow roses, and grabbed a quick kiss on the cheek. Then he quickly snatched her hands as if they were both illegal aliens avoiding capture. While running they ran hand in hand, she had forgotten her luggage at the airport. Well, Bridgett thought, *So much for my sexy and seductive attires.* A white stretch limo located in front of the airport stuck out vividly, and Bridgett was in awe as David gently pushed her into the back seat. When they pulled away, Bridgett kept looking back at the airport. He watched and finally asked, "What's wrong, honey?" Bridgett shrugged her shoulders and answered him saying, "I guess if we go anywhere, I'll have to be naked. All my belongings were left behind." David laughed heartily. "You worry too much, I'm here now to take care of everything; which includes you." He pulled her so close that she was nearly sitting on his lap, and then he started kissing her savagely as if they had been away from each other for a lengthy time. Bridgett heard the statement David made as he pulled her near while holding on tightly, saying, "Soon, you will be mine forever." That statement made her uneasy as it reminded her of Tom, causing a sadness to take over that happy *moment. Maybe*, she thought, *karma had taken precedence over what should have been a very precious time.* Bridgett now fully understood the deep rooted feelings of love and the indepth feelings that were involved. Tom expressed them about her many times without embarrassment and in front of everyone.

Bridgett brushed away the thoughts and feelings of Tom from emerging any further. She was now entering another plane of existence, and this was about questioning David's desire for her without letting him know. A woman thing was the name of the game. What was the phrase? "Oh yes, predator versus lover." Bridgett thought that was a great one. Once the thoughts stopped popping into her head the limo felt as though it was climbing upward while twisting around and around, as if it was encircling some type of path. When the kissing finally ceased, Bridgett looked out of the window, and her guess was legitimate, as the Limo was indeed climbing up a mountain. When the ride finally ended her and David descended from the Limo Bridgett felt completely immersed in a déjà vu experience. She tried to catch her breath and speak, but no voice came out, and she could feel herself starting to panic. Then David's voice was loud and clear, calling her. "Bridgett, are you all right? What's taking place right now? Talk to me!" Bridgett suppressed the thoughts she wanted so badly to explain, but her pulse was racing and her heart was pounding out of control. No voice would emerge. Hearing David's voice brought her back to reality, and she regained composure; which kept the anxiety under control. Bridgett could only surmise what David would have said if told about the déjà vu. She kept hating herself more and more for holding back the realisms that took place in her life to prevent appearing like a helpless, weak and out-of-control freak. Her idea to hide any weaknesses was to put on a little show, laughing gleefully, and kissing David with pretense and fakeness; which were unreal characteristic's of Bridgett. She needed David to believe that everything was wonderful! The returned kisses proved that the scheme worked. In anticipation of his surprise ahead of her, all that was needed would be courage and fortitude. She did answer his question with, "Oh, I was just taking in the view. The climb up the mountain was very unusual, and I was just admiring every part of it." He seemed pleased as he squeezed her hands and replied with great enthusiasm, "Oh, baby, you haven't seen anything yet." And proceeded on with, "Wait until you see the adventure I have kept hidden from you for this special event. It's a very secret place way up in the sky. It will make you speechless! Bridgett stood in bewilderment of what was to follow. The only question that lay present was, "David, how did you ever find this hiding place?" David's reply was, "Oh, I have my ways. I know a lot of influential people." Bridgett was left with no further questions.

What Bridgett witnessed did make her speechless—the endless glass windows that sparkled in the sunlight, the two way glass where people could see out but no one could see in. The scars still remained when relating to "special glass," bringing back the memory of keeping her project safe. Ever since the word "special" came into her life, it always brought dread that linked anything relating to that one word. The glass windows would always be reminders of the lab with its special glass where government officials watched the scientists; but no one could see them. Bridgett hated those glass windows, and a twinge of reluctance occurred upon entering. David kept referring to it as a secret place. However, her thoughts immediately went to secrets, and she avoided them like a virus; because that's what they indicated. Bridgett's plan was to keep everything honest and open in their relationship if it was going to bloom, and believed highly in trust within a commitment. There couldn't be any more hidden secrets. However the honesty aspect with many males, she found, were never brave enough to challenge the words honest and open. Therefore, bonding never occurred with anyone for very long, and she named a couple of the reasons. "Maybe it was her choice not to achieve a relationship." She knew that her personality and characteristics left quite a lot to challenge. The adventure continued with all hopes remaining high that the "secret" surprise would contain a proposal for marriage. However, the night would bring its own. Bridgett became more and more reluctant to know the "secret" surprise. There was no way of relaxing because the déjà vu kept taking place everywhere: making it very difficult to relax. She was going to keep the theme of the adventure, being good or bad, and treat them equally. David told Bridgett on several occasions that she was "his total world, and had never met anyone who could take his soul." He surmised that a lot of his feelings for her came about from never being bored. He had found someone on par with his own intellect, who was so very beautiful and showed a glowing warmth and tenderness. After listening to all she was about, Bridgett questioned herself. Maybe she had never given herself enough credit for being this type of person. She could be very hard on herself, and was her worst enemy. The adventures they shared had always been exciting; however, somehow this adventure was making her a bit edgy. The uneasiness kept flooding inward, and soon she would drown. Bridgett always trusted David with these adventures, and couldn't understand these feelings of uneasiness. He had told her many times, "Bridgett, our relationship has to be based upon truth, alleviating all dishonesty." Keeping that thought in mind, she continued to remain waiting for his surprise as the mirrored secret place reached out for an embrace. She held David's hand, and he squeezed hers in a gesture that said, "Everything will be all right." A small part of Tom emerged again, leaving his mark, as no other male ever squeezed her hand with that particular thought transference. Bridgett wondered if David was as real a person, as she believed him to be, and again he squeezed her hand.

There were no clues as to what to expect. Everything was just going at such a fast pace that Bridgett wondered if she would ever catch-up. This adventure would be labeled as attempts to overpower the inner abilities of a woman. Being his girlfriend of one year yielded the first time feeling uncomfortable with him. Bridgett's feelings were that of defeat; it seemed as though her secrets and fears were already known. Bridgett began to wonder if this was the same David she encountered a year ago that day. The man who was involved with her personal thoughts and emotions. Somehow, she was feeling more as a trophy that he had won for first place—as David dressed her with complete uniformity, to pinpoint perfection from undergarments to cologne going all the way to arranging her hair. Bridgett felt like she was about to play a role in the movie *The Stepford Wives where all the wives looked and acted exactly the way their husbands wanted them.* She wondered if all that was accomplished and prepared by him that evening was nothing more than "inspired aggressions of lust." If so, then there was a seed planted concerning marriage; because everything was going as planned. Bridgett was more than ready to settle down for marriage and children. She wanted this more than being involved any further within the Implantation Program. She would even settle for an engagement ring; because that was a promise ring: which made a commitment. Her mind was racing so fast that too many emotions were surfacing, and everything became a blur as she watched David's mouth moving in conversation, but hearing nothing. She just let him talk as he hooked her by the arm and the long gown trailed behind. They were now headed for the inside of the windowed monolith. He brought his arm around her waist to bring them closer together. All that was needed was a red carpet, and a lightening rod; as both of them together caused lightening to strike. Bridgett watched as heads turned from every direction to gaze upon their entrance. They walked into a large room that appeared to be a lounge and dining area combined. Flashes came from everywhere as cameras began clicking away. They were approached by people neither one knew as they spoke nonsensically while attempting to justify their presence. They both knew this situation all to well, and David winked at her as an added sign of recognition. Their appearances together was no shock to anyone as their pictures were always displayed everywhere. They were even on several magazines and on entertainment segments of television. Introductions to their relationship were always captioned with many different headers like—scientist and senator molds for bizarre couples. They were both alright with it though, and found it to be amusing.

Once Bridgett and David were seated and alone, with the lights dimmed, and candles lit, the room took on a different atmosphere. Soft music filled the room from a pianist near by, and Champagne was poured all adding to the romantic atmosphere. Bridgett had to admit that she was able to relax, and did so by glancing out the nearby windows. Bridgett never imagined how close the sky was to them on top of a mountain. Everything had become a spectacular light show. Out in the distance, the city lights were spread beneath her and above, the sky turned on its many twinkling stars and they became partners and danced. The different spectrum of multicolored lights were like prisms hanging down with each one holding their own difference of brilliance. The mystical display, she believed, was only for her that evening, as this would be a very special evening in her life. Bridgett had never encountered the radiance of so many different colors as they lit up the entire area surrounding them with the twinkling and sparkling of multifaceted gemstones. There was rhythm of the notes being played in the background that were in tune with the light show. She continued sipping on the champagne while entranced by the light brigade. Bridgett had traveled around the globe and could not acknowledge seeing anything that surpassed the array of beauty on display. She found the heavens taking on a form of breathless encounters as it touched the earth. In her heart she became part of them. This event was something to be shared only by a special person, and with that thought in mind, Bridgett turned quickly to David. "Oh my god, David, I'm so sorry." David laughed, staring at Bridgett. "I was soaked into your beauty as you were bathing in the universe. It's okay. Really, I loved watching the different expressions on your face." David squeezed her hands, and she allowed it. He pulled her near as their lips met, and their kiss was that of the sky, being Olimpus as it reached down to Gaia, the earth, which seemed to last forever.

One bottle of champagne emptied and another was ready to be opened as David kept pouring. When the pianist took a rest, the room grew quiet, making it feel that they had the entire room to themselves. Bridgett locked eyes with David as he spoke. "I wanted this night to be special." She knew he was about to begin that something special, and her body became numb and limp as she tried to regain a proper posture. Trying to minimize the excitement that was taking place, she nervously quivered, "I know."

David began, "I wanted it to be so special that you would remember it forever. Happy anniversary, darling." Then they clicked their glasses together.

"God, David, I can't believe you have friends that knew this place existed."

Bridgett remembered the story when he had explained the prominent friends who hooked him up with the secret place. The piano player returned, and David took Bridgett's hand saying, "Let's dance. I want to feel your body next to mine." Bridgett's nervousness had left, and in its place was pure happiness as she took his hand to the dance floor.

However, as David stood up, straightening his clothes, Bridgett noticed something had fallen out of his pocket, and bent down to retrieve it. "Don't!" she heard him yell loudly.

"David," Bridgett said quietly, "calm down. It was right here by my chair. It's not a big deal."

Then, Bridgett handed David the fallen article, and saw what it was he was trying to conceal. It was a gold card with black numbers. It was them; he knew them as well.

David quickly placed the card into his pocket and, with an outstretched hand, waited for hers. Bridgett waisted no time saying, "I don't want to dance right now."

"Why, Bridgett, what's wrong?" David asked quickly.

"Nothing," Bridgett remarked. "I guess I had too much Champagne, and I feel too tipsy."

"Oh, that could be. It's all right, then we'll just sit right here so I can talk with you," he stated.

Bridgett knew damn well that he knew the significance of the card. Yet he was playing her, pretending somehow, someway it didn't happen.

"I have something very important to tell you," he began.

"To tell me or ask me?" Bridgett wanted that clarified.

"To tell you," he answered.

"All right, I'm ready. Tell me." Bridgett said, although she was puzzled as to the wording. Bridgett expounded on those words. Yes, he specifically said, "Tell me." She lost track somewhere; it wasn't making any sense, so she wanted to get to the bottom of the subject and quickly. Again Bridgett said, "Tell me then, I guess."

There would be no escaping what David was about to say, and so the journey began. He continued with his smile; her eyes drifted to his lips. She loved the fullness they presented. His kisses jump started her heart. David continued to stare deep into Bridgett's eyes. She knew something was erupting, and the lava would cause her to melt. She could feel the eruption begin, and it was definitely hot.

"You know why I brought you here Bridgett?" The lava continued to spew forth into the air, far up and hotter than Hattie's. Bridgett didn't want to play anymore. She wanted to hear him propose. She was now completely over the mellowness of the champagne and spoke directly into his face and said calmly, "Yes, I believe so. It was my understanding that it has been a year to date that we have been together, but I'm not sure about anything else. Can you clarify more simply what it is that you wish to tell me because I really need to hear, it David?" But David took it further than Bridgett ever expected. "I love you, Bridgett, more than you'll ever know, and I will always love you. I don't want to be without you in my life. I couldn't bear to think what life would be like without you in it." Bridgett quickly responded without hesitation, "I don't want to be without you either, David." With their eyes still engaged with a hypnotic stare between the two, the volcano blew.

David sat with posture erect. More statuesque then she had ever noted him to be positioned. Bridgett continued to observe his repositioning techniques. The expressions on his face were also reflecting changes. All that Bridgett observed emanating from David was a sign that he was feeling uncomfortable. "David," Bridgett snapped, "out with it for

Christ's sake. I'm running out of patience with you! How difficult can it be?

What is it that you want to tell me? Why is it so hard for you to speak?"

The silence heard was like the menacing arousal occurring just before the advance of the dark dreams. Yes, that was the feeling; Bridgett was sure of it. The dead silence. The cancer. The feeling that is seldom welcomed and usually brings bad news. David was still holding her hand as the environment started to change. The room became icy cold; there was a thickness in the air that felt heavy causing difficulty in breathing. Darkness was starting to surround them. The entire room was immersing into the dark dreams, and she knew what they brought . . . nothing. They filled souls with nothingness, just worthless thoughts that sprang forth dread. Bridgett could recognize them as their black shadows crept out and their outlines were visible. They continued as macrophages. Bridgett had convinced herself that she would allow their attack without a fight as the darkness swirled around her like she was caught in the eye of a hurricane. Every word pronounced bounced back at her from everywhere. "Bridgett, I have to tell you the truth. Maybe you'll respect me more, but I can't hide the secret any longer.-I'm married!"

There, now the dark dreams released their information, and this was the new torment they provided. Anything that was associated with emotional pain and sadness or maybe they really came to provide the truth. This was the first time Bridgett gave herself freely to the dark dreams allowing them to enter-with permission. Now they couldn't hold her hostage; because that's how they played. Once released from the dark dreams without difficulty or aftermaths involved, the momentum and energy that once existed was given back along with all the courage and strengths that were parts of Bridgett Montgomery. Could it be that without the dark dreams that night she would have been nothing more than a pathetic mess. A left over from someone's four-course meal.

She removed her hands, asking for a repeat. "David, what did you just say? I'm sure I didn't hear what I think you just said." David spoke in slow motion, lips parting slightly, enabling the pronouncing of each word to be dragged slowly from his mouth. She finally reached a conclusion that everything surrounding them was in slow motion that evening. Bridgett's rational was that she was drunk, and she wanted desperately to believe that assumption continuing to blame the champagne. Then the words came forward once again as Bridgett dissected each and every syllable attentively as it projected from David's mouth. "I said I need to tell you the truth about something."

Bridgett felt all right with that statement and continued, "That's fine, David, go ahead."

In Bridgett's mind the worst scenario was perhaps that he cheated. But it wasn't, although unlike other women: she wished it had been. David's confession blurted forth with great speed and power.

"I'm married Bridgett, but please hear me out, please." "I didn't want to lose you. That's why I've never told you. I have never loved any woman as deeply as I do you, not in my entire life. I need you in my life, Bridgett, without you I have none." She heard the clarity of it all, remembering that speech once before. Glancing out the windows for the last time that evening, was thoroughly hypnotizing—not David, the magical light show spectacular. The dancing lights were welcoming her to join them. She felt as though they had something in common, as pure happiness emanated giving thanks to the stem cell program that was brought to life. Along with offering their respect and their love for another humans attempt to save humanity. The sky with its uncountable number of stars and with the entire city of lights spread out below, bringing the understanding of the same beliefs they shared. The knowledge of them knowing everything about the stem cell program made her head spin out of control, around and around, playfully at first, then turning into a sickness filled with darkness, and dread, as the dark dreams engulfed her completely—and with permission.

It was no surprise when the light show ended, and the room became silent. Bridgett studied the man across the table and laughed loudly as if they were at a comedy club, and David was the comic on stage. Bridgett wondered if he really believed her to be weak and pathetic to the point of being grateful for his compassion and the expression of love that was entangled with lust. This male species seated before her believed his own theory of how to domesticate a wild animal. Somehow this theory seemed very weak to say the least. Taking a wild animal out of its natural habitat and genetically transferring all it's wild and predatory traits and replacing them with tame docile abilities? Bridgett attacked this probability as it had never been documented, and left no proof. Therefore, there was no foundation for its existence to evolve. She based the statistics by those who died with attempts to domesticate wild animals: which were never accurate, because those who tried died. Therefore they could only be approximated as any and all attempts to erase the innate abilities of wild animals was quite apparent, proved by the wild animal that won. While continuing to stare across the table as a predator toward a prey, Bridgett wanted to show him the innateness of her character provided by an element of a surprise attack. That part of her personality was never shown to David, and lucky for him because it wasn't ladylike and was not a pretty picture. There were actual fears present now, as the smell of it emanated from David. "You pathetic excuse for a Y chromosome that enabled you to become a male has now been jeopardized. In its place is a feminine X chromosome that has proved to be the most powerful breed. How embarrassing is that?" she mused while staring at him. Bridgett finally realized that beauty had to be in the eyes of the beholder because she was about to gag on how unattractive and stupid she had been for letting it go unnoticed all this time. Now that the proof of his deadly deeds was all laid out, Bridgett blamed herself for living the lies David created for their own personal reality.

David kept staring at Bridgett with the expression that he posed quite often as love and admiration. Bridgett totally discredited his species, labeling them weak and vulnerable, while attempting to undermine the true victors—women. David did a thorough job of his homework to meet her; that was positive. Bridgett believed him to be a challenge and was prepared for them as well. However they were disguised as stimulating abilities only, as she predicted. The problem with David was that ultimately he believed he was the winner of all challenges for women. As a man who walked with great statute, his posture was slumping in a downward position, beaten by his own inadequacies to losing the battle which was his ultimate challenge—her. There were beads of sweat covering his face as she smelled his defeat. With Bridgett, there were no games, because she was built on truth and integrity. David knew this from the first day they met at the Hampton Lodge. She knew that he was thoroughly amazed with her beauty and knowledge and matched with the same beliefs and moral issues surrounding the project *he played her*. Once again thinking that he was strong enough mentally to change a feral to be submissive, placed him in a vulnerable position, which brought disaster into his own life.

Her gaze was locked on David's eyes, which she once admired and once set her heart blazing with passion. The more she stared into them, the more she watched as he crumbled into tiny pieces strewn all around. His words were not well pronounced as he faltered with sentences, leaving garbled communication. Everything he was about vanished. In its place was an egocentric and vain apparition of all that entailed the meaning of a "real man," as he humbled himself at her feet. Kneeling before her as if bowing to an idol with apologies, she was sure more lies were added. David made several attempts to stop her from walking away by hanging onto her legs during pleads. Bridgett knew that without her in his life, he would be a source of turmoil, which would lead to his own self-destruction. However, she didn't care what happened to him; the damage was done and irreconcilable. Bridgett wanted him out of her life forever, never saying a word while staring at him. The worst part of a breakup is the silence that can be heard. Bridgett knew that as a rule between genders, this would be his worst punishment. The room was now filled with the presence of many people as they circled around them, with their eyes peering as if their breakup became a free soap opera to view. The display of David's behavior was something to behold and would certainly leave a lasting impression for everyone to remember. Bridgett stood up, pushed the chair away, took the heel of her foot, and pushed it against David's chest to release his grasp and headed for the lobby, leaving David still on his knees, groveling on the floor where he belonged. The story in the Bible regarding Jacob taking his mistress from Babylon and telling her not to look back because that meant she was regretting her decision to leave. He warned her that she would turn into a pillar of stone, and she did. Bridgett felt as if she was living that part of the Bible now, differently though because she had no reason to ever look back at David. She could hear wailing and sobbing coming from some unknown area within the glass windows. Also, there were pathetic cries of someone unknown that could not be recognized; she was sure it was a man who had lied to his lover with stupidity and ignorance that once believed in superiority. Possibly even in the belief that they were the strongest sex, which left the women to be weak was her understanding. He played her and played her right up to the end.

As she woke once more from the dark dreams, Lacy was doing her usual hovering. Bridgett had just visited the sparkling Emerald City of Oz, as defined by the wonderful spectrum of the universe. Bridgett knew she was home as the outline of Lacy and her distinct voice was recognized, giving her Hell. "For God's sake, Bridgett, you're scaring the shit out of me. I should never have gone along with your scheme. Stop it, Bridgett, stop it! Come back. Please come back." Those pleads echoed through Bridgett's mind. "Wake up, please, Bridgett, wake up." Multiple attempts were made to react to the words which echoed from within the blackness of the cage where she was being held hostage. Bridgett noticed that the harder she fought to free herself this time, the stronger the dark dreams had become while keeping their position, and holding tightly. Bridgett fought harder than ever before, now wondering if this was the last hurrah. The distinct smell of freshly dug dirt was a new odor detected. It felt as though it was being thrown upon her body indicating she was dead and was about to be buried alive. This belief was way too hard to comprehend and could not possibly be happening. She wondered if the dark dreams were through playing and was disposing of her body. Bridgett's thoughts raced to the reality of which was very surreal. The dirt continued to be thrown upon her, mostly directed to the facial area preventing her from breathing. If she screamed, dirt would be consumed, and suffocation would be undeniable. Suffocation was known to be the worst way in which to die, and it was said to be a slow process. Bridgett knew that panicking would only make the torment harder to endure. Her mind continued racing with ways to be released. With extremities still free, she took advantage of that aspect quickly before the shadow figures noticed. She threw her hands as far up as they would go and started digging herself from out of the grave. It made sense now, as thoughts brought about why embalming became a law. This gave a prime example as Bridgett proved it to be the worst way to die. Trying not to panic wasn't an easy effort, as the inability to breathe became a reality. Screaming was useless, as they were muffled by all the dirt thrown upon her body; no one could hear the screams. Bridgett continued fighting desperately by digging toward the top, as it was too cramped along the sides and given no room to move. Yes, the top was a better choice, as she threw them up and up to the top, hoping someone above would see them, and she'd be rescued. Bridgett could feel the cold dampness of the earth being sucked into her body as she shook and trembled from the damp and cold, combined with the never-ending darkness. Bridgett waited for the next event while continuing to dig upward. Then with the process of being buried alive, suffocation taking place, and the coldness of a corpse, other things emerged from the darkness. Bugs, she knew they were bugs as Bridgett felt them biting into her flesh. *My god!* Her thoughts raced. The bugs were eating her alive; they were feeding off what appeared to them as a dead and decomposing body. They were snipping at her flesh, and there wasn't any space to get away from them. They kept filling their little bodies with flesh nibbling and nibbling, while the heaviness of all the dirt was still being thrown upon her became so very heavy that she could no longer throw her hands upward out of the grave anymore. Suffocation was inevitable, as breathing became shallow. Soon Bridgett's lifelessness would become a real corpse, allowing all the suffering to finally end. Then, the only salvation would be death while being buried six feet under, as every part of her memory was shown on a giant screen. Noises were being heard from somewhere. First began the mumblings followed by garbled words. Bridgett concentrated and listened intently as some of the sounds were somewhat clear. She continued with attempts to understand the words. Now a voice was recognizable. The voice continued to be persistent as it became louder, which then turned into screams. At first, Bridgett thought it was the last of her brain participating in life, as it began to die, leaving a trail of earthly hallucinations still attached to the world where she once lived. Bridgett even saw the light that was often depicted by those who died and came back to tell their story. The light became more visible, glaring, causing her vision to be impaired. She believed it to be her final demise.

Then the vision of an angel took place, as envisioned by her personal beloved friend and colleague, Lacy. As Bridgett's eyes opened, Lacy's tears dripped softly down upon her face like morning dew, unnoticed but felt. It was Lacy's devotion that forced Bridgett to replace death by the will to live. Whatever she did that brought her home would encompass a strong love and bonding of souls. "Thank you, Lacy, I guess I owe you." Bridgett smiled. Lacy stroked her hair softly.

"Bridgett, your body feels like it's on fire! God!" Lacy yelled. "Why are you so red hot? Even my hands are burning when I touch you. Look, my skin is peeling off." Bridgett gazed up at Lacy as her head still lay within her hands and said, "Maybe I just returned from Hell!" Then Bridgett stayed perfectly still for a countless period of time to understand that a humans will to survive was a lot stronger than anyone realized. What really transpired within the dark dreams was the strength to live and the ability to hate. This wall of hatred that transpired between her and the dark dreams became a vengeance as she grasped any and all information, before waking, that was fundamental to their existence which would cause their surrender.

Arriving at the airport, Lacy demanded a wheelchair and assistance to take Bridgett home. Lacy stayed with Bridgett once they arrived inside their safety zone. Lacy made some tea and sat with Bridgett on the couch. "Bridgett"—Lacy said her name quietly—"did you remember anything from the dark dreams? Are you sorry about allowing them to take you? What was it you found out? You owe me. You said you did. Please tell me." Lacy's pleads were accepted as being in debt, so Bridgett told everything that happened through that last journey with the dark dreams. Everything was told to Lacy about David, their meeting, his lies, along with deceits, which gave birth to the bizarre light brigade that did not end as surmised by being buried alive. She gently poured out everything to lighten her burden. All the hidden secrets that were built as a fortress to keep everyone away, protecting her sanity from being viewed by many as crazy. Bridgett had never felt so free and happy as she did at that moment in time. *No more secrets, no more lies. It was a feeling that could not be expressed by any human*, she thought. She even explained to Lacy the excitement that filled the room as she left David still kneeling upon the floor. Lacy was stunned at all the information that was withheld all this time. This gave her a better understanding of the pain surrounding Bridgett, which released both of them from their tormenting afflictions. Lacy still remained amazed as the information of the dark dreams continued with impeccable details.

Bridgett knew it was overwhelming information that would take Lacy a huge amount of time to accept not only as the truth, but in the realization of all the turmoil that had taken place in their lives, as in the end it turned full circle; which meant freedom. Bridgett sat and absorbed all of Lacy's behaviors and viewed them in detail as the significance related to processing everything. Bridgett felt an energy surge returning to her soul, the peacefulness that would now allow her to rest. Lacy had never shared a sad or tearful episode with anyone, as her gift just allowed to only feel emotions. Since Bridgett had built a wall to her secrets, she could not understand them. Now Lacy could connect all the pieces of the puzzle that belonged to Bridgett's character and allowed her once again to gain composure of her faculties and pride. Lacy could do nothing more but gaze at Bridgett in awe of all that transpired from the beginning of the dark dreams to the end of the relationship with David. Tears now flowed uncontrollably, shedding all the emotions connected with their breakup and the pain of the dramatic ending. The way it ended was simply the worst scenario she had ever encountered. Lacy wondered how Bridgett would deal with this dramatic end between her and David. Lacy knew that she could now become a better friend and confidant to Bridgett, as she earned her respect. Bridgett seemed to read Lacy's thoughts as she brushed away all the tears and smiled, saying, "Men don't make women weak. They just make them a hell of a lot stronger."

The days that passed were unaccountable to Bridgett. Her thoughts ensured that the pieces of puzzle were now tightly fitted together. She had allowed the dark dreams to pull her in without a fight. Bridgett could now remember David and the year they spent together along with the memories of the ugly hidden truth that spoiled everything that was once beautiful. *Maybe,* Bridgett thought, *I am what Tom says—gullible—for believing that everyone tells the truth, because there's nothing that horrible to hide.* She embraced the terrible mourning of her lost love as the days passed. The despair of what had once been and was now gone was bittersweet. She called Tom about business concerning the labs, and he gave her assurance that everything was good and then asked her out for dinner. Bridgett declined for a few good reasons. She didn't know how much he knew about her affair with David. Something that Tom would find quite obvious, because he'd read her eyes and in doing so would see the pain she felt for another man. She would be an easy map for Tom to follow and see the enduring loss of David. The pain that she would try to conceal from the world had left a hole where her heart once belonged. Then Bridgett would see Tom's pain of losing her, and it would provide nothing but guilt. She didn't want to deal with anyone or anybody right now. She just felt like being alone with her thoughts. Anything. Bridgett felt that she needed more time to heal and took a rest from the conferences and passed them on to Lacy. She was thrilled and eager to do them.

The scars she wore would never leave, and Bridgett would have to return to life and not just existence. Many women were hurt by men as she and carried the weight of their relationships around hindering the ability to regain their health mentally and physically. The apartment continued to remind her of that hurt, and just brought her feelings down realizing that it was not a safety zone anymore. Bridgett knew that moving on meant forward which entailed finding another place to live. For now, though, she needed a place to think and heal allowing those processes to take place before making any drastic decisions. Her mind knew what she needed and wanted—the lodge. She wanted to be somewhere she felt safe, away from prying eyes and an onslaught of questions. Bridgett knew they would protect her there. With that final decision, she packed to go to the lodge. As the TV blasted away in the living room, someone was making headline news, and she heard a familiar name being repeated. She ran to see if she knew the person. To Bridgett's horror, it was David. She watched him scurrying at a fast pace, like the bugs in the dark dreams as they scurried to devour her as she was being buried alive. He was trying to avoid the public by holding a paper next to his face. But Bridgett already knew the truth with all he represented: a cold, an unconcerned greedy person that was a liar and a cheat. Bridgett could feel herself shaking as her skin once again felt cold and clammy. It was fear that Bridgett was feeling, and yet she couldn't acknowledge it's relation to seeing David's face. It was over, and she didn't care about him anymore. There was no connection related to him. As the news continued and Bridgett tried to understand what was taking place, she read the caption under his picture, "Senator's wife commits suicide at their estate in New Hampshire." This caption would now be embedded forever in her mind, along with the other scars she already carried.

Bridgett could feel herself shut down. While preparing to leave for the lodge, she was thankful that David's wife was never shown. It lessened the quilt and there was no face to remember. However a great factor remained for Bridgett and that was the connection between her and David which questioned the motive of his wife's suicide. Her primary focus at this time however was her sanity; which she would retrieve from staying at the Lodge. She felt numb with no thoughts and was vacant of any feelings. As she sat in the back of the cab, there evolved thoughts of nothing, as her mind was emptied. She just stared out the window and watched everything fly by and couldn't decipher any of it. As the cab rounded up and up through the pines, Bridgett began to smell the aroma of security and safety. She watched the trees as they waved back and forth, as if they were expecting her and were welcoming the arrival. Freedom was close at hand, and she was accepted with open arms by the hospitality of the lodge. Their warm sincere smiles of welcome reminded her of home. The staff continued to greet her with respect by using "Ms. Montgomery," and the key to her private room was given without hesitation. Bridgett was unable to speak; her words were empty as well as her thoughts. All she could do was nod her head in an affirmative manner. Bridgett remained frightened and was very light-headed and dizzy. She became silent. She didn't know if the dark dreams would take advantage of the situation and come for her. She saw something in the staff's faces, but without asking, she would just continue to guess what they offered. They didn't seem to be alarmed; their mannerisms were unchanged. If they knew something, they didn't show it. Her eyes did stop at a newspaper lying upon the desk. She slid it carefully under her arm to read once inside her room. Bridgett tore into the newspaper the second the staff left her room. Nothing was on the front page, nothing in small print or in the obituaries. Knowing that he was the senator of that area, she wondered why he wasn't front page news. Bridgett couldn't stop thinking about David's wife suicide. She mourned her death as if they had been best friend. Being at the lodge allowed her the privacy to grieve about everything since the beginning of the project. As she recalled David's face on the news, she noted absent emotions. He was never seen sad or unhappy or crying; he was just cold and expressionless. His uncaring attitude, fleeing from the cameras and trying to hide. *He never cared about her*, Bridgett thought. He never even had feelings for his own wife. What if they had married? Would he be the same way after her demise? Expressionless. To think how his wife loved him so much that she carried the burden of their relationship to her grave. She couldn't stand the pain of losing him or being without him. *How much she must have loved him*, Bridgett thought, *to take her own life.* How could someone love that much, so hard, so deeply, that they would rather be dead than deal with losing that person to someone else? Bridgett could not think anymore. Her thoughts were becoming too involved with David's life, and she hated thinking about him. She just wanted to escape, even if that meant sleeping. She was above caring about the dark dreams. Her entire life now was torture; the dark dreams would only be a relief. As she quickly fell asleep, her brain kept reminding her of how glad she was that they never showed his wife's picture. Then the last thought before sleep finally arrived was the staff at the lodge greeting her with their voices in unison, saying, "Welcome home, Ms. Montgomery."

Bridgett didn't remember or think about how long she had stayed at the lodge and didn't care. She was safe and happy, and that was all it was about. She had called to alert the staff that she did not want to be bothered, and they did so. *I guess*, Bridgett thought, *since I always told them to hold all calls and not let anyone know I was staying there, they continued to keep it that way. God, they were great people.* Bridgett had all her meals brought to the room. Freshly cut flowers adorned her room daily. No one bothered her, and the phone never rang. She wasn't expected to talk to anyone, including the staff, and it just felt like a little piece of heaven. Bridgett knew that eventually she would have to call Lacy, Tom and her mom, as they would fear for her safety, and notify the police as a missing person. Bridgett wasn't really insane, as all thoughts about that word kept submersing. "I'm just going to enjoy life." was how she understood herself. She never turned on the TV, and always had her curtains wide open to enjoy the leaves changing into different colors. Bridgett felt she had picked the best time to go crazy and run to the edge. She would just sit and stare out at the beauty of nature and inhale its glory as it crowned the earth with its presence. She could still smell the scent of the pines through her window and through her room. The rains came and went, and she could smell the aroma it brought to the pines as they peacefully swayed in the calmness that made her feel safe and secure. There was a oneness with nature that bonded tightly, and it felt so wonderful. Bridgett thought about people in her life but felt no bond with anyone except for the beauty of what was surrounding her at that time. Bridgett did know something for sure and kept reassuring herself of it; that there were never any dark dreams at the lodge. Maybe they knew she was protected, safe, and not vulnerable; and she therefore couldn't be threatened. Bridgett could sleep forever in peace and continued to enjoy the warmth and security she felt being—home.

As the days continued to pass, she felt her strength come back full force. The characteristics that were once a part of her returned. The new Bridgett Montgomery sprang to life with exuberance and fortitude, as she reclaimed her sanity. Bridgett felt as if the entire planet was now back to its original position and all the inhabitants were also. She would continue with her conferences, open more labs, meet with Tom and Lacy, and have some fun. That's what Bridgett wanted to happen. It wasn't about her sanity, people, or the dark dreams. It was about never living, just existing. Now that she had gotten some well-deserved rest and started a proper diet and the negative ions were inducing many positive thoughts, she was ready to leave the lodge. Her sleep patterns were now back to normal, and she had no further episodes of the dark dreams. Everything came about in a positive direction. Yes, indeed, she couldn't have asked for a better life. She had blown the dark dreams out of their secret existence as Bridgett found the answer to their survival. It was plain and simple fear. It lived and thrived off the fear they induced and what the fear represented. Now it was crystal clear. When they came and held her hostage, all they needed was to be accepted without fear and to leave a piece of them that had to be removed in order to escape with her life. Once that piece was taken from the dark dreams, they could pose no further threat, because they wanted that piece returned before it could be analyzed. They despised any interference with their games. They allowed anything she wanted to dream and produced it as being real for that one piece she took, and she was punished by them for taking it. They wanted it returned. Everything about them was absolutely real in her life, and the struggle needed to happen as they fought for their own existence as well. If a piece of them was given back then the dark dreams would release anyone who entered. They had no control over how she would piece the puzzles together. But she played their game, placing their pieces together effectively and had won her existence to live. Any truth that was found by her was released as part of their game strategy. If Bridgett found the truth she was seeking, they released her. Bridgett was thrilled at herself for fighting back at the dark dreams and not allowing them to destroy her life. She felt the removal of them meant more than winning the Nobel Prize. In proving the theory of the dark dreams, with no regard to fear or concerns, she curled up in her big comfortable bed and went to sleep.

Now that Bridgett had regained her energy, she would also regain herself—her sanity. She practically ran down to the lobby to find the staff that had nurtured her back to health. The old Bridgett was gone, and the new one arrived. She was very anxious for everyone to notice the great changes that had taken place while turning in her private key before leaving. Her self-respect and dignity had been regained. It had been lost during the torrid relationship. Bridgett approached each staff member with a hug, and they responded in return using their warm and affectionate smiles; applying no forms of fakeness to be noticed. Bridgett laid the key to her private room number twenty-four upon the desk. She was ready to face life again and the new adventures it would bring. Then, a hand slid in and picked it back up. Bridgett turned to see the person that shared the same private room, and the perpetrator that took it made her come face-to-face with the devil incarnate. David stood before her, staring deep into her eyes, and the only response was to lock in the stare, and no words were exchanged. Bridgett felt brave enough to speak first. Her need for closer was necessary. Thoughts began racing around in her head, thinking of the right approach, and all that submerged was, "Hi, David, how are you?" After some moments of speechlessness, her words began to flow easier. "I'm sorry about your wife. Is there anything I can do to help you through this trauma? I mean as a friend." David unlocked his gaze as the key to room number twenty-four remained in his grasp and finally spoke. "I appreciate your condolences, Ms. Montgomery." Hearing his voice caused Bridgett intense anxiety. The room temperature dropped dramatically, and an icy coldness caused her to shudder and a cold wind appeared out of no where surrounding them. She felt like a corpse kept on ice. But, Bridgett stood her ground as her body reacted to the fight-or-flight syndrome produced by fear, and what it would unveil. David's voice was frightening, and left no understanding as to the reason. It wasn't the voice that Bridgett recognized as belonging to him. Bridgett tried to conduct herself in a calm demeanor, as he responded, "I do appreciate and thank you for investing your valuable time on me." Bridgett did not like the snappy manner. "David!" Bridgett said his name very harshly. "David, if you want to be hateful, you're doing a good job. However, you're leaving the door wide open for me to do the same, and I don't think you want me to go there, do you? Therefore, I will replace that hatefulness by asking a sincere question that has been ripping my inner soul apart." "Wow, Ms. Montgomery," David said very arrogantly, "I'm impressed that a question could possibly interfere with your soul: I can't wait to hear it." Then Bridgett brought out the big guns and blatantly asked without remorse, "My question, David Bradford, is if your wife's suicide was connected to our relationship? I need to know the truth so I can move on with my life, and you can do the same." She watched as David's eyes interlocked with hers even deeper. Bridgett heard him quite clearly, as well as the others; who were becoming spectators to their conversation. It was a Hell of a show that day when the devil and Bridgett collided. He began with, "Dear Ms. Montgomery, surely you have more to do in your world conferences than to speak one-on-one without a crowd. Your expertise in the world of science with stem cells and embryonic implantations, I hear, is a smashing success. You have opened many throughout the country. Bravo for you! Myself, I would love to attend all your conferences just to admire your beauty and that little sexy body. It would be such a thrill and cheaper than renting a porno. We could even share room-twenty-four here at the lodge. Really, Ms Montgomery, you must have a lot of groupies following you around like a rock star. However, I am not a part of your breed. This is the first time I have actually seen you in person, and I don't know why I waited so long. You're well worth looking at, and then some." "David!" Bridgett's voice rose. "Why do you keep denying our relationship? Why are you pretending that we never happened? I think you're quite vain to have surmised that I never mattered in your life before today, and the love we shared was hardly something to shrug away as you do so well in front of the cameras." David once again turned to Bridgett with eyes blazing red with fury, as the battle with the devil transpired. The coal dark eyes that she once loved, and brought many favorable emotions; were now objects of menacing desires, as portrayed by the dark dreams. Still gazing, with eyes penetrating deeply within her soul, David spoke, "For the books, Ms. Montgomery, there was never a relationship between us. Although I still stand behind my offer for the night. I would give my soul for one more night of love with you. This would also be to your advantage as a reminder of the only man that ever made you light up like a Christmas tree. Now, as to your question regarding my wife and her suicide, I guess will remain a mystery. I was always faithful to my wife, Ms Montgomery. A relationship with you would be like eating your favorite food and never feeling full. Then I'd pay the price for my desire, as the cobra part of you would strike out with vengeance and seep venom deep inside me. This would then force me to always remember you when I looked at my scar." Bridgett allowed David to keep talking. "I'm just speaking the truth. Isn't that rated high in your book of who's who? It is a virtue that you regard highly, isn't it? Truth is high priority, isn't it, Bridgett?" She caught it. The devil slipped, and she got him good, when he said "One more night, and called her Bridgett. He was mortal after all, Bridgett believed. She remained speechless for the first time in her life. A good comeback was always easy for her; now it was difficult. All Bridgett could surmise was, *"It was David, for sure, bringing everything up from the depths of hell to try and take me back with him. He was a master at trickery and deceit, and could disguise his voice and mannerisms as the calm before a storm. She was a challenge to him, and he did his homework very well. So, there were no surprise that he knew what buttons to push.* She also knew there was anger involved with his sweet words; because he would never get a second chance with her. Bridgett believed that he was able to reconstruct all his abilities and hook another women; but it wouldn't be her. She would never reunite with him. Bridgett had to give him credit though, he was very suave and a natural at game playing. David enjoyed adventures that were challenging. She believed she was his biggest challenge, and would never find another to match. Bridgett's fury used up all her ammo, and there was only the eight ball left to play. Desperate actions called for desperate measures. She used the staff for support, and was determined not to go off the deep edge.

Nothing seemed settled between them especially the denial of his extramarital affair, Bridgett snapped. She couldn't stand one more lie coming out of his filthy mouth, and started screaming in the lobby again. I can't believe you're all afraid to put him in his place and tell the truth. Tell him!" she demanded to the staff. "Can't you see he's playing with me and toying with all of you?" Bridgett expounded more upon David's character. "He loves adventures and this is one right now. He lives to play games with other people's lives and doesn't know when to stop even if it means suicide. He never cared about his wife. It doesn't even bother him to know that she committed suicide over an affair that was silently taken to her grave. Listen to me, he's crazy." She had finally gotten everything off her chest. "Checkmate, David. I just stoled your king, were Bridgett's thoughts." He said nothing to prove his innocence only wanting to inflict more pain. Bridgett tried to hold back from physically striking him. She couldn't comprehend where everything was leading. The hatred for him was now engaged into a personal conflict, and she was losing ground. David made her this way, as he brought forth these aggressions of behavior. Bridgett became a woman scorned and had lost her sense of judgment and objectiveness. Hatred consumed everything. Seeing his true colors made her realize she had lost nothing when the relationship ended. Nothing ever existed between them that could ever be considered a loss. David made sure of it by erasing all their memories of happiness and leaving sadness in return.This helped Bridgett immensely to realize the impact she had upon him, and he loathed her for knowing that reason. She felt proud of the accomplishment. Now, she would no longer have to fear his name: spoken in her presence. *Nothing but betrayal and no love involved for him at all*, she found herself repeating over and over in her mind. Bridgett's hate mounted as he mocked her, as a joke for everyone to have a good laugh. The way she left him at the secret place, on his knees as a spectacle for everyone somehow hit the magazines, as this was his state, and he was the senator. With a large crowd gathering; Bridgett stopped caring about how she was perceived. Her need was justification for what was taking place at the lodge, and she was going to get it. "I'm not as crazy as you think," Bridgett blurted out. "Your wife was crazy for killing herself over a pathetic excuse for a man." She felt that nailed him good, and striking over his comment regarding her sanity. Now, she began using the staff in seeking their help. "For God's sake, someone tell him tell him what a liar he is for denying our relationship. Everyone here knows the truth. Why are you letting him get away with these lies? Force him to plead his guilt, and to stop denying the relationship, and all of you can prove that it existed. Make him tell the truth, about us, and our lives together here at the lodge. Please! Please!" Bridgett kept pleading with the staff, searching every face in the crowd for some acknowledgment to come forth. "Tell him you know the truth about room twenty-four. However, the lodge took on a new personality of its own entering into a dead zone; where no one appeared to be real anymore. It changed into something different; yet almost familiar. Bridgett felt a blanket of thick heaviness that brought an icy chill to her bones, and bringing with it familiar feelings of despair and destitution. The darkness invited itself inside as it crashed the party acting like a star performer. Creeping and hiding within a cloud of fog and mist used as a disguise. Bridgett knew this intruder, this unwanted guest, as its characteristics were all too familiar. The darkness unfolded it's self into a vacuum, a void, a black hole, and the abyss as she watched in horror at its vicious attack upon everyone that day at the Lodge: including David. It brought justification to the absolute nothingness that was believed to be reality, as it confronted all barriers of intellect—that turned into dead silence!

The sun shone brightly on that day in July as Bridgett and her husband, Tom, took a stroll with their two-year-old daughter, Britney. *San Diego is a great place to live*, Bridgett thought. *Ideal weather all year-round, temperature hardly changes.* She looked over at Tom, and a smile crossed her face. *I love him so much*, she thought, and then as if in response to her thoughts, he squeezed her hand.

God, it was such a gorgeous day!

CITED WORKS

"International Society Stem Cell Research Ethics, of Human Embryonic Stem Cell Research, Louis Guenin, ISSCR have proof to those who wish to inquire into the moral debate of Embryonic Stem Cell Implantations.

The Ethics of Human Embryonic Research by Louis Guerrin, Feb.2, 2005 has a quote by Edmund F. Pellegrino M.D. of Georgetown University stating a Catholic case against embryonic use as" one who holds that we should treat every embryo as a person for purposes of the "not to kill," embryo-destructive experiments could gain justification only if it were argued within any moral view." Catholic Church against opposition into the use of embryos purposes of "the duty not to kill."

Argument of Kevin Fitgerald, Ph. D. From Lolyo university, Biochemists, his opposition is that "Biologists can't draw a line somewhere after fertilization that marks the start of Human Life!" He believes that "embryos deserve the same protection as infants and shouldn't be destroyed to obtain cells."

Ann Drakyain, and Ruth Faden science article from McGal Hills, willing to donate frozen embryos for scientific research.

Margaret Farley, Ph. D. of Yale University explains her moral optimistic view of Embryonic Stem Cell Research, also found within the "Ethics of Human Embryonic Research," by Louis Guerrin, "is more than one catholic's thought of reasoning." She concludes in her theory of stem cell research as "one who recognizes humanitarian use to finally end suffering."

Barack Obama, President of the United States in his article from the EVSCO, by Saloly, Scientific America, June 2000 addresses and decides that Embryonic Stem Cells will be funded from the United States Government along with appointing several scientists to powers of authority within the Executive Branch.

www.ingramcontent.com/pod-product-compliance
Lightning Source LLC
Chambersburg PA
CBHW030847180526
45163CB00004B/1485